David C. Brenner

Handling of Bulk Solids

Theory and Practice

P. A. Shamlou

Department of Chemical and Biochemical Engineering
University College
London

Butterworths

London   Boston   Singapore   Sydney   Toronto   Wellington

All rights reserved. No part of this publication may be reproduced
or transmitted in any form or by any means, including
photocopying and recording, without the written permission of
the copyright holder, application for which should be addressed to
the Publishers. Such written permission must also be obtained
before any part of this publication is stored in a retrieval system of
any nature.

This book is sold subject to the Standard Conditions of Sale of
Net Books and may not be re-sold in the UK below the net price
given by the Publishers in their current price list.

First published 1988

Reprinted 1990

© **Butterworth & Co. (Publishers) Ltd, 1988**

British Library Cataloguing in Publication Data

Shamlou, P. A.
   Handling of bulk solids.
   1. Bulk solid materials. Materials handling
   I. Title
   621.8'6

   ISBN 0-407-01180-3

Library of Congress Cataloging-in-Publication Data
   Handling of bulk solids.
   Bibliography: p.
   Includes index.
   1. Bulk solids handling. I. Title.
   TS180.8.B8S56 1988   621.8'6   88-4307
   ISBN 0-407-01180-3

Photoset by TecSet Ltd, Wallington, Surrey
Printed and bound in Great Britain by Courier International Ltd, Tiptree,
Essex

Preface

Of all the materials handled and produced by the process industries, the greatest bulk is in the solid state and almost always in particulate form. In the chemical industry alone the value of product formed as particles is greater than 30% of the whole. The handling of particles is big business often done wastefully from an energy point of view, and improvements in techniques could lead to considerable savings over a wide range of industries. Compared with our understanding of how to handle, move and store gases and liquids, our understanding of particles in bulk is rather primitive.

The subject is difficult and multi-faceted. Particles are rarely handled in a vacuum, and interaction between particles and intervening gas or liquid plays an important part in overall behaviour. Despite, or probably because of, the complex nature of particulate systems, there are relatively few books published covering the field. In particular, the few books that are currently available on the market are either highly specialized, appealing to only a limited group, or too superficial, forming a small section of a much more general book. What is needed is a book that covers the whole field of solids flow and handling in the process industries — a book which treats the subject in depth and is suitable for advanced undergraduate and postgraduate levels as well as for practitioners in industry.

This book is intended to fill this gap between experienced researchers and those new to the work. The subject matter presented is difficult to obtain elsewhere without referring to many sources of information, so the author has assumed a role of interpreter of the literature with those findings and equations being presented which can be of immediate utility to the reader.

Presentation of the subject follows classical lines of separate discussions for each topic, so each chapter is self-contained and can be read on its own. However, a background in mathematics at the first-degree level and an appreciation of the concept of Mohr circles is helpful for a proper understanding of the material, particularly in Chapters 2 and 3.

Worked-out examples are included at the end of each chapter to familiarize the reader with the numerical manipulations and orders of magnitude of various parameters which occur in the subject of bulk solids handling. Because of the complicated form of most of the design equations involved, the computer is an ideal vehicle for the solution of many design problems in bulk solids handling. Indeed computer-generated solutions have been utilized throughout this book, but the author has resisted the temptation to include specific computer programs because each computer installation is slightly different in its input-output capability and, in any case, such programs are not difficult to write and a number of these are readily available on the market.

The field of bulk solid handling is not static. New developments occur regularly, as depicted by the huge amount of information which is available in the open literature. For that reason, each chapter also includes a comprehensive and up-to-date list of references which can be used by the reader to build up further investigation.

I am grateful to Professor P. N. Rowe for suggesting that I should organize a course at University College London on the subject of bulk solids handling. This subsequently formed the foundation for many of the ensuing chapters.

I also wish to express my gratitude to Professor J. W. Mullin for first seeding the concept of this book in my mind and then helping me to materialize it by his useful comments and advice on the organization and layout of the chapters.

I am also indebted to many of my students and colleagues who contributed generously to the preparation of this book. Special thanks are due to Dr John G. Yates for so efficiently reading the final manuscript and for his useful suggestions. I should also like to thank Keyvan M. Djamarani for diligently checking the equations, computer programs and numerical examples.

My thanks are also due to Miss Mary P. Bell for so patiently and skilfully reading the whole manuscript and correcting my English.

P. A. Shamlou

Dedication

To my wife Monir and my children Naby and Nieki for their forbearance during the writing of this book

Contents

1	**Bulk solids flow and handling properties**	1
	Introduction	1
	Particle and bulk properties	2
	Particle-particle and particle-fluid interactions	9
	Bulk solid handling and flow behaviour	12
	References	17
	Symbols	17
	Example	18
2	**Pressure profiles in bulk solids storage vessels**	20
	Introduction	20
	Variation of pressure with position in a particulate under storage	25
	References	45
	Symbols	46
	Example	47
3	**Design of storage vessels for particulate solids**	51
	Introduction	51
	Mass-flow silos: flow/no flow criterion	52
	Funnel-flow silos	73
	References	75
	Symbols	76
	Example	77
4	**Gravity flow of particulate solids**	80
	Introduction	80
	Gravity discharge of coarse particles	84
	Gravity discharge of fine particles	91
	References	99
	Symbols	100
5	**Pneumatic conveying of bulk solids**	102
	Introduction	102
	Positive-pressure pneumatic conveying systems	104
	Dilute-phase pneumatic conveying systems	108

	Dense-phase pneumatic conveying systems	118
	References	122
	Symbols	123
	Example	124
6	**Hydraulic transport of particulate solids**	130
	Introduction	130
	Settling suspensions	132
	Non-settling slurries	138
	Equipment components	145
	References	147
	Symbols	148
	Example 1. Settling suspensions	149
	Example 2. Non-settling slurries	153
7	**Mechanical conveyers**	156
	Introduction	156
	Screw conveyors and elevators	157
	Belt conveyors	161
	Bucket elevators	163
	En masse conveyors	166
	Other conveyors and feeders	167
	References	168
8	**Safety in bulk solids handling**	169
	Introduction	169
	Dust fires and explosions	169
	Health hazards	179
	Dust control equipment	180
	References	188
	Index	189

Chapter 1

Bulk solids flow and handling properties

Introduction

Enormous quantities of bulk solid materials are produced by the process industries each year; in the chemical industry alone the value of product formed as particles is greater than 30% of the whole. In any operation involving particulate solids, successful storage, flow and handling of the bulk material is a major and essential part of the overall plant design.

The proper design of bulk solid storage and handling equipment, in turn, requires knowledge of the individual and bulk properties of the particulate material under static and dynamic conditions. There are many such properties including the following (Anon., 1970):

- particle size, shape, size distribution and surface area;
- particle and bulk density;
- cohesive properties, flowability and fluidizability;
- hardness, compressibility;
- toxicity, flammability and explosibility;
- optical, thermal, magnetic and chemical characteristics;
- hygroscopicity (ability to attract moisture).

The relative importance of these properties depends largely upon the particular unit operation under consideration (Table 1.1). Many of the characteristics listed above, e.g. those relating to flammability, explosibility and toxicity, are secondary properties; consequently, their definitions and methods of measurement are often highly empirical, requiring considerable expertise to obtain and interpret meaningful data.

The aim of this chapter is to give an account of some of the more important solids properties that influence the behaviour of bulk materials during handling operations. The treatment is confined to those properties for which a fairly satisfactory basis exists for the interpretation of experimental results.

2 Bulk solids flow and handling properties

Table 1.1 Importance of particle and bulk properties to some solid handling operations

Unit operation	Important bulk and particle properties
Storage and gravity discharge from bins, silos and hoppers	Size, size distribution, shape, particle and bulk density, cohesive and frictional properties, fluidizability, flowability, explosibility, toxicity and compressibility
Pneumatic and mechanical conveying	Size, size distribution and shape, particle and bulk density, friability, toxicity and explosibility
Hydraulic conveying	Size, size distribution, particle density, friability and dispersibility

Particle and bulk properties

Particle size, shape and surface area

Particle size, shape and surface area are fundamental characteristics of bulk solids and are of paramount importance in most unit operations involving such materials; these properties are closely related and should be considered together. They determine to a very large extent the degree of interaction of particles with the surrounding fluid and with each other. These interactions, in turn, influence critically the behaviour of the bulk material, e.g. its flowability, fluidizability, compressibility, toxicity, flammability and explosibility. Unfortunately the relationships between these basic parameters and the practical behaviour of bulk materials are not yet fully understood.

Table 1.2 Classification of bulk solid materials according to size

Size range (μm)	Standard term		Characteristic
	component	bulk	
30 000–3000 (but may be as low as 1000 μm)	grain and lump	broken solid	free-flowing, but could cause mechanical arching problems during discharge from bins and silos
1000–100	granule	granular solid	easy-flowing with cohesive effects if % of fines is high
< 100	particle	powder	
(i) 100–10	particle	granular powder	may show cohesive effects and some handling problems
(ii) 10–1	particle	superfine powder	highly cohesive; very difficult to handle
(iii) < 1	particle	ultrafine powder	extremely difficult (or impossible) to handle

Bulk solids flow and handling properties 3

In general, no universally accepted method has yet emerged to define and classify particles according to their grades. Table 1.2 lists some of the common terms relating to particle size. For a spherical particle, the size is defined uniquely by its diameter. However, with the exception of a few powders, the shapes of most industrial particles are irregular and the definition of particle size presents some difficulty. To overcome this, size is sometimes expressed in terms of the diameter of a sphere equivalent to some property of the particle. Table 1.3 lists some of the more common equivalent diameters and their definitions.

Particle size may also be expressed in terms of a statistical diameter; typical examples are Feret's diameter and Martin's diameter (Allen, 1981).

The more irregular the particle, the greater is the variation between the various equivalent diameters. Therefore, the shape of the particle is equally important and needs to be specified also. Particle shape may be defined in several ways. One approach is to define the sphericity, ψ, of the particle as:

$$\psi = \frac{\text{surface area of a sphere having the same volume as the particle}}{\text{surface area of the particle}}$$

that is

$$\psi = \left[\frac{d_v}{d_s}\right]^2 \tag{1.1}$$

where d_v is the diameter of a sphere with the same volume as the particle and d_s is the diameter of a sphere with the same surface as the particle.

Table 1.3 Equivalent diameters of irregular particles

Equivalent diameter	Definition
Projected area d_p	Diameter of a circle with the same projected area of the particle when viewed in a direction perpendicular to its most stable position ($A = \pi/4 d_p^2$)
Volume d_v	Diameter of a sphere with the same volume as the particle ($V = \pi/6 d_v^3$)
Surface d_s	Diameter of a sphere having the same surface as the particle ($S = \pi d_s^2$)
Sieve d_a	The width of the minimum square opening through which the particle will pass
Specific surface d_{sv}	$d_{sv} = d_v^3/d_s^2$
Free-fall diameter d_f	Diameter of a sphere with the same terminal velocity and density of the particle
Drag $d_d \simeq d_s$	Diameter of a sphere with the same resistance to motion as the particle in a fluid of the same viscosity and having the same velocity
Stokes' d_{st}	$d_{st} = d_v^3/d_d$

4 Bulk solids flow and handling properties

For a spherical particle, $\psi = 1$. All factors remaining the same as ψ deviates from unity, the particles become less spherical and consequently less flowable (more cohesive) and more difficult to handle.

Equation 1.1 may be used to obtain values of the sphericity for geometrically regular bodies such as cubes and spheres; it is perhaps worth noting that for most such forms, ψ takes an average value of 0.77 with a standard deviation of $\pm 11\%$ (Geldart 1986). For industrial solids, while ψ can have a value as low as 0.28 for a material such as mica flakes, for the majority of powders its value falls in the range of 0.65 to 0.98 (Geldart 1986). Unfortunately, in practice it is not generally easy to determine values of ψ for irregular particles because of experimental difficulties associated with the determination of d_v.

Heywood (1963) devised an alternative method using the equivalent projected area diamater, d_p, as the basis to define the equivalent surface area diameter and the equivalent volume diameter of the particle. Thus:

$$\text{particle surface area} = f d_p^2 = \pi d_s^2 \tag{1.2}$$

that is

$$f = \frac{\pi d_s^2}{d_p^2} \tag{1.2a}$$

$$\text{Particle volume} = k d_p^3 = \pi d_v^3 / 6 \tag{1.3}$$

that is

$$k = \frac{\pi d_v^3}{6 d_p^3} \tag{1.3a}$$

where f and k are the surface and volume coefficients respectively and are determined experimentally (Heywood, 1963). The value of f/k (i.e. $d_s^2 d_p / d_v^3$) is a measure of the shape of the particle; it represents the influence of shape upon specific surface.

Heywood's method of definition of particle shape assumes that the particle is in its most stable position. If the particles are randomly orientated, as they will be in practice, then different values of f and k will be obtained. However, except for excessively elongated particles, the definition provides a reasonable estimate of the shape. In this respect it is worth noting that if, instead of the projected area diameter, the equivalent volume diameter, d_v, is used to assess the shape of the particle, then k remains constant (equal to $\pi/6$) and f will be independent of orientation. Unfortunately, d_v is difficult to obtain experimentally, and thus d_p is preferred, since it can be determined relatively easily for all sizes of particles.

Heywood also observed that if a particle has the dimension T, B and L in order of increasing magnitude, then the following ratios may be defined:

$$\text{Elongation} = n = L/B \tag{1.4}$$

and

$$\text{Flatness} = m = B/T \tag{1.5}$$

Bulk solids flow and handling properties 5

For an equidimensional particle, that is $B = L = T$ and $n = m = 1$, the volume coefficient may be described by the following relationship:

$$k_e = km \sqrt{n} \tag{1.6}$$

k_e has unique values for geometrical bodies such as cubes and spheres, and values may be assigned with a reasonable degree of accuracy to particles which approach one of these (Heywood, 1963).

The surface coefficient, f, is more difficult to obtain. Heywood provided the following empirical equation for the estimation of f based on many measurements on large particles for which the surface area was determined directly:

$$f = 1.57 + C \left[\frac{k_e}{m} \right]^{1.333} \left[\frac{n+1}{n} \right] \tag{1.7}$$

where values of the constant C and factor k_e are given in Table 1.4 for various geometrical forms.

Although it is often possible to assign a mean shape coefficient statistically to a sample of bulk material, most industrial particles when tested show considerable variation in shape. For coal dust, Heywood (1961) provides some data for the distribution of both k and k_e. His results indicate that about one-third of the variation in k is due to variation in geometrical form, the other two-thirds being due to variation in the proportions.

It is also possible to express sphericity, ψ, in terms of f and k by noting that

$$\psi = \pi \left(\frac{6k}{\pi} \right)^{2/3} \left(\frac{d_p^2}{f d_p^2} \right) \tag{1.8}$$

$$= \pi^{1/3} 6^{2/3} \frac{k^{2/3}}{f} \tag{1.9}$$

thus

$$\psi = 4.836 \frac{k^{2/3}}{f} \tag{1.10}$$

Table 1.4 Heywood's shape coefficients

Shape	k_e	C
Approximate bodies		
rounded	0.54	2.1
subangular	0.51	2.6
angular, tetrahedral,	0.38	3.3
prismoidal	0.47	3.0
Exact shapes		
spherical	0.524	1.86
cubic	0.696	2.55
tetrahedral	0.328	4.36

6 Bulk solids flow and handling properties

Particle mean size and size distribution

Most industrial particle-forming and processing operations result in particles of every size, up to a maximum which is usually fixed by some controlling process. If a sufficiently large sample of this material is graded artificially according to size, the results may be presented either on a mass basis or on a number basis, depending upon the method of measurement of particle size. In each case, the data may be tabulated or plotted in various ways, e.g. as frequency (or relative frequency) histograms and polygons, percentage distributions, and cumulative distributions.

Attempts have also been made to describe the grading information mathematically by fitting empirical equations to experimental data. Numerous expressions have been proposed to describe the spread of particle sizes including the Gaussian (normal), log-normal, and Rosin–Rammler (Table 1.5).

An example is shown in Figure 1.1 in which the distribution of cement particles is plotted against varying size ranges on semi-log paper; the data were obtained using standard dry sieving techniques. The results may also be described conveniently by the Rosin–Rammler distribution law, i.e.

$$R = 100\, e^{(-bd_p^{n'})} \tag{1.11}$$

where R = volume of material left on a specific screen
  e = cumulative % of coarse grains
  d_p = particle size
n' and b are distribution parameters obtained from the data.

Table 1.5 Various distribution functions

Type of distribution	Expression
Normal (Gaussian)	$Ni = \dfrac{\Sigma N_i}{\xi\sqrt{2\pi}} \exp\left[\dfrac{-\left(d_i - \dfrac{\Sigma d_i N_i}{\Sigma N_i}\right)}{2\xi^2}\right]$
Log normal	$\xi = \sqrt{\dfrac{\Sigma N_i\left(d_i - \dfrac{\Sigma d_i N_i}{\Sigma N_i}\right)^2}{\Sigma N_i}}$ $N_i = \dfrac{\Sigma N_i}{\log\xi_g \sqrt{2\pi}} \exp\left[\dfrac{-(\log d_i - \log d_g)^2}{2(\log \xi_g)^2}\right]$ $d_g = \sqrt[N_i]{d_{i1}d_{i2}d_{i3}\ldots\ldots d_{N_i}}$ $\log \xi_g = \sqrt{\dfrac{\Sigma N_i(\log d_i - \log d_g)^2}{\Sigma N_i}}$
Rosin–Rammler	$N_i = -bn(d_i)^{n-1}\exp(-bd_i^n)$

Bulk solids flow and handling properties 7

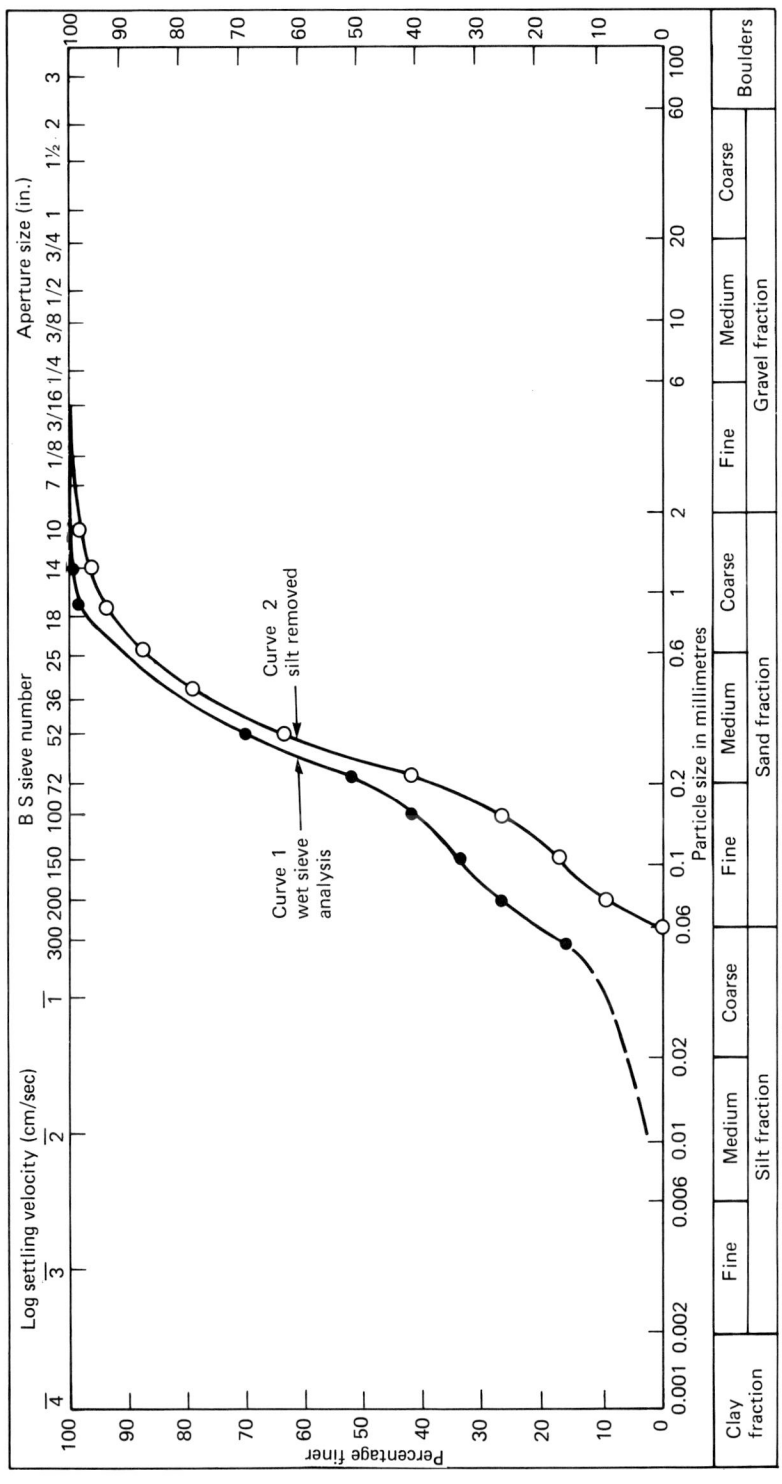

Figure 1.1 Particle size distribution for cement.

8 Bulk solids flow and handling properties

Equation 1.11 suggests that a plot of log log (100/R) against log d_p (or log (100/R) vs d_p on log–log graph paper) should yield a straight line of slope n'. The constant b may be evaluated by substitution into Equation 1.11.

It is also useful to be able to describe the size of a group of particles by an average value representative of the group. Table 1.6 lists some of the mean diameters that have been used for this purpose. In practical applications, the specific definition used should relate closely to the process under investigation; for example, in operations involving mass and heat transfer, the surface area of the particles is an important parameter and thus an appropriate diameter to use in such cases would be the surface mean diameter.

Particle size measurement

Measurement of particle size should not present any specific problem. The subject has been studied extensively and reviewed thoroughly elsewhere (Allen, 1981).

Essentially, particle size may be obtained by measuring the size of each individual particle. This may be achieved conveniently by either scanning a field of view of the particles or by counting the particles falling through a sensing device in single file. Numerous techniques and commercial equipment are available to achieve this and Table 1.7 lists some of the more common methods. There is clearly some overlap between the various methods listed in Table 1.7. In general, different methods of measuring a given parameter for apparently the same material do not necessarily yield the same result. This is not to say that one method is better than another, but merely reflects the fact that the various techniques measure the same

Table 1.6 Various definitions of mean diameter for particles with a size distribution

Average size	Expressions
Mean surface diameter	$d_{ps} = \left(\dfrac{\Sigma N_i d_i^2}{\Sigma N_i} \right)^{1/2}$ or $\left(\dfrac{\Sigma x_i/d_i}{\Sigma x_i/d_i^3} \right)^{1/2}$
Weight (volume) mean diameter	$d_{pw} = \dfrac{\Sigma N_i d_i^4}{\Sigma N_i d_i^3}$ or $\Sigma x_i d_i$
Surface-volume (Sauter) diameter	$d_{pvs} = \dfrac{\Sigma N_i d_i^3}{\Sigma N_i d_i^2}$ or $\dfrac{1}{\Sigma x_i d_i}$
Numeric mean diameter	$d_{pl} = \dfrac{\Sigma N_i d_i}{\Sigma N_i}$ or $\dfrac{\Sigma x_i/d_i^2}{\Sigma x_i/d_i^3}$
Linear mean diameter	$\dfrac{\Sigma N_i d_i^2}{\Sigma N_i d_i}$ or $\dfrac{\Sigma x_i/d_i}{\Sigma x_i/d_i^2}$

To convert from number density to mass density it is assumed that particle density remains constant. Thus: $x_i = \rho_p N_i k d_i$

quantity on a different scale of 'size' (Rose, 1961); indeed the different results might be all correct. Therefore, in specifying a value for a certain parameter, it is important to quote the means by which the parameter has been measured. Moreover, the method of measurement should be closely related to the application of the final product; for example, for catalyst specification, an appropriate mean size characterizing the particles may be obtained using permeability and gas adsorption, while optical microscopy may be used effectively in order to assess the covering power of a paint pigment.

Particle-particle and particle-fluid interactions

The surface area of a powdered material increases enormously with decrease in particle size as illustrated graphically in Figure 1.2 in which a reduction in particle size by one-eighth is shown to lead to an increase in the surface area by 800% (Green, 1975).

With the increase in surface area, the magnitude of surface forces on the particles also increases. Such forces may arise from electrostatic, van der Waals and capillary effects. The magnitude, and hence the significance of these forces, depend critically on particle size and are also affected strongly by moisture content, surface roughness and impurities, particle shape, and humidity.

The resulting forces between the particles may be either attractive or repulsive, depending upon the nature of the surface forces. When a high degree of attractive (cohesive) forces exists between the individual particles, the powder will become less free-flowing, and consequently more difficult to store, discharge, fluidize and convey.

Table 1.7 Common methods of particle size measurement

Particle size range (μm)	Principle technique and equipment	Parameter measured
0.001–15 0.8–800	Field scanning: electron microscope optical microscope	Projected area length and statistical diameter
1–2000 0.5–50	Stream scanning: Coulter counter light scattering	Surface volume diameter
1–100 0.05–25 0.05–50	Sedimentation gravity centrifugal X-ray	Free-fall diameter drag diameter projected area
1–40 40–3000 > 3000	Screening wet sieving dry sieving grizzlies and grids	Sieve diameter
0.005–50 0.1–50	Surface area: gas adsorption permeametry	Surface volume diameter surface diameter

10 Bulk solids flow and handling properties

Electrostatic forces

Electrostatic forces existing between pairs of particles or particles and surfaces are due to either a surplus or deficit of electrons. Electrostatic charging may be induced intentionally during powder processing or incidentally during handling operations. Electrostatic charging due to frictional contact could occur from movement of surfaces over each other, either powder against powder or powder against a wall.

The magnitude of electrostatic charges is strongly affected by a number of variables including particle size and shape, surface roughness and impurities, humidity, and moisture content. For non-conducting particles, the distribution of charges on the surface of the particle is also likely to be non-uniform. As a result, it is generally not easy to estimate the magnitude of electrostatic forces acting between pairs of particles and particles and surfaces. For two charged particles, the following equation may be used to estimate the electrostatic force experienced by each particle:

$$F_{el} = -\frac{Q_1 Q_2}{\xi 8.86 \times 10^{-12} Z} \tag{1.12}$$

where F_{el} is the force in newtons, Z is the centre-to-centre distance between the particles in metres and ξ is the permittivity of the interstitial fluid ($\xi = 8.86 \times 10^{-12}$ A s/V m for air). Q_1 and Q_2 are the charges on particles in coulombs. For dispersed fly-ash particles below 10 μm, Clift (1985) suggests that the average charge per particle may be obtained from the following empirical relationship:

$$Q = -3.973 \times 10^{-11} (d_v)^{1.025} \tag{1.13}$$

Thus, for a particle of 5 μm in size, equation 1.13 results in a charge of about -1.47×10^{-16} C. This indicates a surplus of about 920 electrons on the surface, noting that the charge per electron is 1.60206×10^{-19} C.

The electrostatic force of attraction between a metallic spherical particle and a polished metal surface may be obtained using the following equation developed by Russell (Klinzing, 1981):

$$F_{el} = \frac{Q^2}{16\pi\xi \left[0.5772 + \frac{1}{2} \ln \frac{2R}{Z} \right]^2 RZ} \tag{1.14}$$

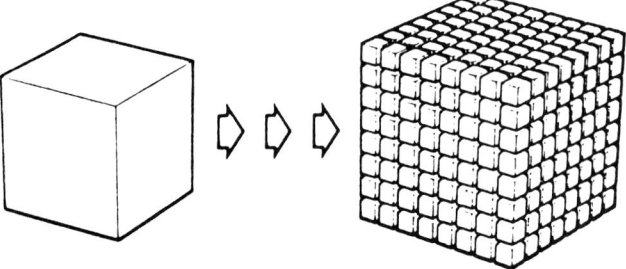

Figure 1.2 Increase in surface area due to reduction in size (Green, 1975).

where Z is the particle–surface spacing and R is the radius of the particle. From Equation 1.14, the force of attraction between the 5 μm particle and a conducting surface at a particle–surface separation of about 4 Å will be about 2×10^{-19} N. Harnby et al. (1985) point out that most reported electrostatic forces of attraction are in the range 10^4–10^7 N/m². The magnitude of the electrostatic force has a second-order dependency with respect to Q, and decreases linearly as particle–particle and particle–surface spacing decreases (Equations 1.12 and 1.14).

Van der Waals' forces

Van der Waals' forces of attraction occur on a molecular level and are essentially due to variations in the local electric fields contained in the solid body; these electric fields in turn originate from the continuous polarization of the atoms or molecules within the body.

Lifshitz found that the van der Waals' force of attraction between a sphere and a flat wall can be determined using the following expression:

$$F_{van} = \frac{\hbar\omega R}{8\pi Z^2} \tag{1.15}$$

where Z is particle–surface spacing ($\simeq 4$ Å), R is the radius of the particle and $\hbar\omega$ is the Lifshitz–van der Waals constant with a value which depends upon particle properties.

For van der Waals' force of attraction between two sperical particles, a similar equation has been proposed (Orr, 1966):

$$F_{van} = \frac{\hbar\omega}{8\pi Z^2} \left[\frac{R_1 R_2}{R_1 + R_2} \right] \tag{1.16}$$

Taking the 5 μm particle from the previous example and the same particle–surface spacing of 4 Å and assuming an average value of Lifshitz–van der Waals constant of about 10×10^{-19} J will give a van der Waals force of attraction of about 6×10^{-7} N.

Theoretical van der Waals interparticle forces per unit area range from 2×10^7 to 3×10^8 N/m² (Rumpf, 1962; Harnby et al., 1985); thus for particles that have relatively large areas in very close surface contact ($\simeq 4$ Å), van der Waals forces of attraction are significantly larger than forces arising from electrostatic effects.

Capillary forces

As moisture content increases, electrostatic forces become less important, while attractive forces arising from physically adsorbed liquid films and liquid bridging predominate; at low humidities the force of attraction is mainly due to the overlapping of the adsorbed film layers between adjacent particles and adhering surfaces, while above a critical value of humidity, bonding is primarily due to liquid bridges between the particles (Harnby et al., 1985).

12   Bulk solids flow and handling properties

For relative humidities approaching 100%, the force of attraction between a particle in contact with a surface may be estimated using the following expression (Klinzing, 1981):

$$F_{cap} = 4\pi\eta R \cos \varphi \qquad (1.17)$$

where φ is the wetting angle and η is the interfacial tension. To give an example, Equation 1.17 results in an attractive force of about 1×10^{-6} N for a 5 μm particle and water on a smooth surface with $\varphi = 65°$ and $\eta = 73$ mN/m.

For two particles in contact and a fully wetting-liquid Clift (1985) recommends the following equation developed by Fisher for pendular bridges:

$$F_{cap} = \frac{\pi\eta d_p}{1 + \tan\dfrac{\delta}{2}} \qquad (1.18)$$

where δ is the angle defining the size of the bridge and approaches zero as particle–particle spacing decreases; thus as $\delta \to 0$ the capillary force of attraction between the particles will approach $d_p\pi\eta$.

For an agglomerate with a circumference C and cross-sectional area A, and containing uniform spherical particles with diameter d, the tensile strength of the agglomerate in the capillary state (T) may be obtained using the following equation due to Rumpf (1962):

$$T = \left[\frac{6}{d}\left(\frac{1+\epsilon}{\epsilon}\right) + \frac{C}{A}\right]\eta \cos \varphi \qquad (1.19)$$

Capillary forces of attraction are substantially higher than those due to electrostatic and van der Waals' effects.

Bulk solid handling and flow behaviour

Despite the considerable work that has been done in developing fundamental theories and measuring techniques for interparticle forces, the relationships between these forces and the practical behaviour of the bulk material are still not well understood; the literature is full of examples of materials that are all seemingly cohesive and yet show considerable variation in their flow and handling behaviour. To overcome such difficulties a number of empirical and semi-empirical approaches have been proposed. These are considered below.

One method of categorizing powders that has found wide acceptance is the empirical classification proposed by Geldart (1973) and reproduced in Figure 1.3. This gives an indication of whether, and if so under what circumstances, a given powder can be fluidized by a gas. According to their fluidization behaviour, powders are classified into groups A, B, C and D.

Group B powders include materials with a mean particle size in the range $40 < d_p < 500$ μm and density ranging between 4 g/cm^3 and 1.4 g/cm^3. Group B powders are easily fluidized, forming bubbles with all gas velocities only slightly above the minimum fluidization values.

Group A powders also fluidize easily, but expand considerably before showing any sign of bubbling. Even then, bubble formation is rather limited.

In contrast, powders in group C are highly cohesive and difficult or impossible to fluidize; the gas merely passes through channels extending from the distributor to the surface of the pile of the powder in the bed. The behaviour of group C powders is often attributed to the presence of strong interparticle forces (van der Waals, electrostatic, adsorbed moisture, and capillary); the magnitude of these surface forces is substantially in excess of that due to drag of the fluidizing gas which is exerted on the individual particles. As particle size decreases, the relative magnitude of the interparticle forces decreases in comparison with the drag force. With group A and B materials, the relative magnitude of the drag force exceeds that due to interparticle attraction and, as a result, group A and B powders are easily fluidized (Molerus, 1982).

By allowing for the effect of interparticle forces Molerus (1982) provided a basis for interpreting Geldart's empirical classification diagram. His analysis, which involved van der Waals forces of attraction, was later extended by Seville and Clift (1984) to include interparticle capillary forces. Figure 1.4 shows Molerus's explanation of Geldart's classification; evidently the transition from group A to group C behaviour is not sharp, and depends upon factors such as shape and relative surface hardness and roughness. The transition between group A and group C becomes wider (the dotted line in Figure 1.4) in the presence of other forces due to electrostatic forces and moisture content.

Molerus's interpretation of Geldart's classification is remarkably good, particularly for the transition between group A and C behaviour, considering that the curve drawn by Geldart is purely empirical and noting some of the uncertainties associated with the estimation of interparticle forces (Schubert, 1984).

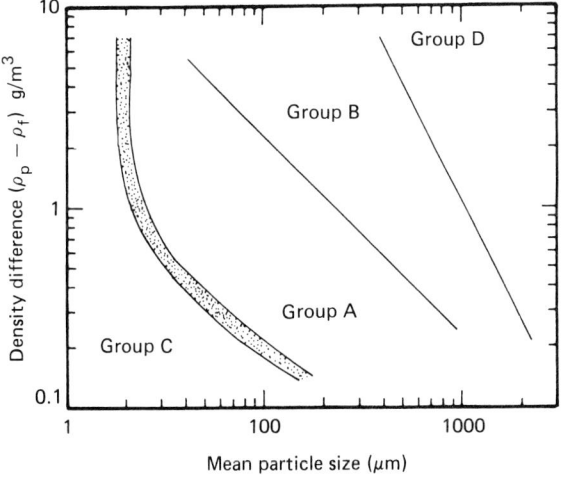

Figure 1.3 Geldart's powder classification diagram for fluidization by air (Geldart, 1973).

14 Bulk solids flow and handling properties

Recently Foscolo and Gibilaro (1984) have provided an alternative and more plausible explanation of the fluidization behaviour of powders by gases and liquids. They point out that while interparticle forces are certainly present and influence powder behaviour in bulk form, their effect upon the behaviour of most fluidized suspensions under equilibrium condition is questionable. These workers have provided an alternative interpretation of Geldart's classification which is based solely on the fluid–particle interaction forces of gravity, buoyancy, and drag. Moreover, Gibilaro and Di Felice (1985) also argue that interparticle forces, if present at all, will only influence the dynamic behaviour of the fluidized bed; in such cases they suggest that the effect of interparticle forces can be accounted for easily by the inclusion of an appropriate term in their hydrodynamic model.

The hydrodynamic model of Foscolo and Gibilaro has been tested successfully against experimental data over a wide range of variables including changes in pressure, temperature, and gravity (Gibilaro and Di Felice, 1985). It has provided a theoretical basis for the transition between Group A and Group B behaviour in Geldart's diagram. The model is currently being extended to describe the boundary between Group A and Group C powders (Gibilaro et al. 1987).

An alternative engineering solution sometimes adopted in practice is to develop the so-called 'yield locus' diagram for the material experimentally using, for example, the Jenike shear cell apparatus described in Chapter 3. The yield locus is a plot of the critical shear stress at which flow occurs against normal (compressive) stress acting on the sample during testing. The plot uniquely characterizes the bulk material for a given set of conditions, e.g. preconsolidated stress and time, moisture content, and particle properties such as size, size distribution and shape; for the same bulk material, a change in any of these factors will result in a different characteristic curve of the type shown in Figure 1.5.

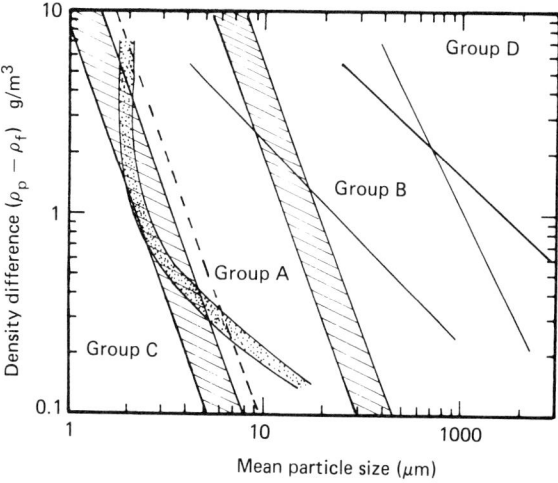

Figure 1.4 Molerus' interpretation of Geldart's powder classification diagram (Molerus, 1982).

For the sake of the present discussion, it is perhaps interesting to note the general form of the experimental plot in Figure 1.5. For an ideal free-flowing powder with little or no interparticle forces, the plot is usually a straight line going through the origin (line A). In the presence of strong interparticle forces the line intercepts the shear stress axis (line B) at a critical point (S); the value of S is a measure of the apparent cohesion of the powder and gives an indication of the stress required to cause flow in a powder under no compressive load. Moreover, the slope of the line is a measure of the angle of internal friction of the powder, μ. A mathematical equation describing the experimental yield locus of an ideal cohesive bulk solid may be written in the following form:

$$S = S_0 + \mu p \tag{1.20}$$

In practice, most industrial bulk solid materials produce a characteristic curve which is of the type shown by curve C in Figure 1.5 (Jenike, 1962).

Two further important parameters may be extracted from shear test results; these are the so-called unconfined yield stress, f_c, and the major consolidating stress, $\bar{p}_1$. These factors are obtained experimentally using a graphical procedure described in Chapter 3. For the purpose of this presentation, it is worth noting that f_c represents the strength of the material at a free surface and is obtained by drawing the Mohr circle with its centre on the compressive stress axis passing through the origin and tangential to the material yield locus. Keeping all other factors the same, the value of f_c depends upon the magnitude of the (major) consolidating stress, $\bar{p}_1$, present during sample preparation. $\bar{p}_1$ is located by drawing the Mohr circle with its centre on the compressive stress axis and tangential to the end-point on the material yield locus; the end-point is obtained experimentally and represents the condition at which flow occurs with no change in volume of the specimen.

The ratio of the major consolidating stress to unconfined yield stress ($\bar{p}_1/f_c$) is a powder characteristic that determines its flowability and,

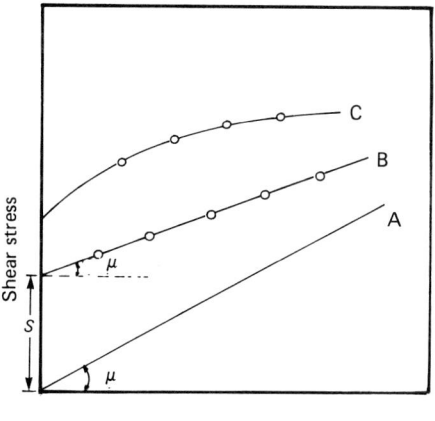

Figure 1.5 Material yield loci for ideal and industrial bulk solids.

16 Bulk solids flow and handling properties

indirectly, the degree of cohesion (Table 1.8). A mass of fine particles with a large surface area and a high bulk density gives a yield locus curve with a high angle of internal friction, a high apparent cohesion, S_0, and a low value of $\bar{p}_1/f_c$. Consequently, the bulk material will exhibit considerable non-flow characteristics and could present problems during handling and transfer operations. On the other hand, a mass of large particles with a low bulk density and a small surface area generally results in low angles of internal friction, low apparent cohesion and a relatively high ratio of $\bar{p}_1/f_c$. As a result, this bulk material will be easy-flowing, and provided it is also chemically stable should not present undue problems during subsequent handling, storage and conveying operations.

Bulk density

The bulk density of a particulate material is defined as the weight per unit volume of a large group of the particles. Measurement of bulk density should not present any specific problem, and several standard methods have been proposed for its evaluation (Carr, 1965a,b). The difficulty is in interpreting the results from such measurements, since several definitions of bulk solids exist, including the aerated (loose), packed, tapped, fluid, average (mean of aerated and packed), and the working bulk density, which Carr (1965) defined as:

$$\rho_w = (\rho_p - \rho_A)C + \rho_A \tag{1.21}$$

where ρ_w, ρ_p and ρ_A are the working, packed and aerated bulk densities respectively, and C is compressibility expressed as a fraction defined as:

$$C = \frac{\text{packed bulk density} - \text{aerated bulk density}}{\text{packed bulk density}}$$

Given the present state of knowledge, it is difficult if not impossible to establish any relationship between the bulk density of a material and its flow behaviour: bulk materials with widely different bulk densities can have remarkably similar flow characteristics. Nevertheless the bulk density of the material being handled is needed for many purposes, including the evaluation of the capacity of bins and hoppers, the estimation of bulk solid discharge from silos, and the calculation of the compressive strength of the material.

Table 1.8 Jenike's classification of flow behaviour of bulk solids

Value of $\bar{p}_1/f_c$	Nature of bulk material
> 10	free-flowing
4–10	slightly-flowing
2–4	cohesive
< 2	very cohesive; non-flowing

References

ALLEN, T. (1981). *Particle Size Measurement*. Powder Technology Series, 3rd edn. London: Chapman and Hall
ANON. (1970). *Classification and Definitions of Bulk Materials*. Book No. 550, Rockville, Maryland: Conveyor Equipment Manufacturers Association
BAILEY, A. G. (1984). *Powder Technol.*, **37**, 71–85
CARR, L. JR. (1965a). *Chem. Eng.* (Jan. 18), 163–168
CARR, L. JR. (1965b). *Chem. Eng.* (Feb. 1), 69–72
CLIFT, R. (1985). *Powtech 85* Inst. Chem. Eng. Symposium Series No. **91**, Birmingham, UK. pp 27–41
FOSCOLO, P. U. and GIBILARO, L. G. (1984). *Chem. Eng. Sci.*, **39**, 1667–1675
GELDART, D. (1973). *Powder Technol.*, **7**, 258–292
GELDART, D. (ed.) (1986). *Gas Fluidization Technology*. Chichester: Wiley, Interscience.
GIBILARO, L. G., DI FELICE, R. and FOSCOLO, P. U. (1987). *Powder Technol*. In Press
GIBILARO, L. G. and DI FELICE, R. (1985). *Chem. Eng. Sci.*, **41**, 2438–2440
GREEN, K. D. (1975). *Powder Technol.*, Series **6**, 76–83
HARNBY, N., EDWARDS, M. F. and NIENOW, A. W. (eds.) (1985). *Mixing in the Process Industries*. London: Butterworth
HEYWOOD, H. (1961). In: *Powders in Industry*. SCI Monograph No. **14**. Society of Chemical Industry. pp 146–149
HEYWOOD, H. (1963). *J. Pharm. Pharmacol.* [Suppl.], **15**, 56T
JENIKE, A. W. (1962). *Trans. Inst. Chem. Eng.*, **40**, 264–271
KLINZING, G. (1981). *Gas–Solid Transport*. New York: McGraw-Hill
MOLERUS, O. (1982). *Powder Technol.*, **33**, 81–87
ORR, C. JR. (1966). *Particulate Technology*. London: Macmillan
ROSE, H. E. (1961). In: *Powders in Industry*. SCI Monograph No. **14**. Society of Chemical Industry. pp 130–149
SCHUBERT, H. (1984). *Powder Technol.*, **37**, 105–116
SEVILLE, J. P. K. and CLIFT, R. (1984). *Powder Technol.*, **37**, 117–129
RUMPF, H. (1962). In: *Agglomeration*, Knepper, W. A. (ed.). Interscience, N.Y. 379–418

Symbols

A	agglomerate cross-sectional area and particle projected area
B	particle breadth
b	constant in Equation 1.11
C	agglomerate circumference and constant in Equations 1.7 and 1.21
d	particle diameter
d_i	particle diameter in the size interval i
d_a	particle equivalent sieve diameter
d_d	particle equivalent drag diameter
d_f	particle equivalent free-fall diameter
d_p	particle equivalent projected area diameter
d_s	particle equivalent surface diameter
d_{sv}	particle equivalent surface volume diameter
d_v	particle equivalent volume diameter
d_{st}	particle equivalent Stokes' diameter
d_{pl}	particles length mean diameter
d_{ps}	particles surface mean diameter

18   Bulk solids flow and handling properties

d_{pv}	particles volume mean diameter
d_{pvs}	particles volume–surface mean diameter
d_{pw}	particles weight mean diameter
F_{cap}	capillary force of attraction
F_{el}	electrostatic force of attraction
F_{cap}	van der Waals force of attraction
f	surface coefficient
$h\bar{\omega}$	Lifshitz–van der Waals constant
k	volume coefficient
k_e	parameter defined by Equation 1.6
L	particle length
m	flatness
n	elongation
n'	constant in Equation 1.11
N_i	number of particles in the size interval i
Q	particle surface charge
R	cumulative % of (coarse) grains (Equation 1.11) and particle radius
p	normal stress
$\bar{p}_1$	major consolidating stress
S	shear stress and particle surface area
S_0	apparent cohesion
T	particle thickness and agglomerate tensile strength
V	particle volume
x_1	mass fraction of the particles in the size interval i
Z	particle–particle or particle–surface spacing
ϵ	voidage
ζ	standard deviation
η	interfacial tension
μ	angle of internal friction
ξ	permittivity
φ	wetting angle
ψ	sphericity
ρ_p	packed density
ρ_A	aerated density
ρ_w	working density

Example

Particle size analysis on a 100 g sample of a mixture of polymer powder gave the following sieve data:

Mean size (μm)	850	710	600	500	425	355	300	250	212	180
Mass retained (g)	3	17	19	15	12	10	7	7	5	5

Bulk solids flow and handling properties 19

Determine:
(i) the numeric mean size
(ii) the surface–volume (Sauter) mean size
(iii) the weight mean size
(iv) linear mean diameter
(v) the specific surface assuming a volume coefficient, $k = 0.21$ surface coefficient, $f = 3.0$ and mean particle density $\rho_p = 1300$ kg/m^3

Solution

The various mean sizes are given in Table 1.6.
Table 1.9 shows all the parameters that need to be calculated first:

(i) the numeric mean diameter $= \dfrac{\Sigma x_i/d_i^2}{\Sigma x_i/d_i^3} = \dfrac{757 \times 10^{-8}}{2.83 \times 10^{-8}} = 268$ μm

(ii) the Sauter mean diameter $= \dfrac{1}{\Sigma x_i/d_i} = \dfrac{1}{249 \times 10^{-5}} = 402$ μm

(iii) the weight mean diameter $= \Sigma x_i d_i = 484$ μm

(iv) the linear mean diameter $= \dfrac{\Sigma x_i/d_i}{\Sigma x_i/d_i^2} = \dfrac{249 \times 10^{-5}}{757 \times 10^{-8}} = 329$ μm

(v) the specific surface of the powder is $= \dfrac{f}{k}\dfrac{1}{\rho_p}\dfrac{1}{d_{pvs}}$

$= \dfrac{3.0}{0.21} \times \dfrac{1}{1300} \times \dfrac{1}{402}$

$= 27$ m^2/kg

Table 1.9

d_i μm	x_i	d_i^2	d_i^3	$\dfrac{x_i}{d_i^2}$	$\dfrac{x_i}{d_i^3}$	$\dfrac{x_i}{d_i}$	$x_i d_i$
850	.03	722 500	6.1 × 10^8	5 × 10^{-8}	.06 × 10^{-8}	4 × 10^{-5}	30
710	.17	504 100	3.6	33	.05	24	119
600	.19	360 000	2.2	54	.09	32	116
500	.15	250 000	1.3	60	.12	30	75
425	.12	180 625	.77	65	.15	28	50
355	.10	126 025	.45	76	.21	28	34
300	.07	90 000	.27	80	.27	23	22
250	.07	62 500	.16	110	.43	28	17
212	.05	44 944	.09	120	.60	24	12
180	.05	32 400	.06	154	.85	28	9
				$\Sigma 757 \times 10^{-8}$	$\Sigma 2.83 \times 10^{-8}$	$\Sigma 249 \times 10^{-5}$	$\Sigma 484$

Chapter 2
Pressure profiles in bulk solids storage vessels

Introduction

Throughout the process industries, deep bins and silos are used extensively for the storage and transfer of bulk solids such as chemicals, foodstuffs, pharmaceuticals, cement, coal, polymer, and powdered metals. In most cases it is important to ensure an even mass discharge of the particulate between process stages, with the minimum of alteration in the quality of the stored product. However, hoppers have also found numerous applications as process equipment in which the maintenance of optimum processing conditions to affect physical and chemical changes often involves the removal or addition of heat and the establishment of controlled temperature gradients within the vessel during the handling operation.

The size of bins used in practice varies widely with diameters ranging from less than 1 m for the production of some high-speciality products to well over 30 m for the stockpiling of coal, cement and some food products. Similarly, the range of materials stored can vary widely from dry, coarse, free flowing granular solids to damp, fine cohesive powders with a high degree of adhesion and compaction under storage conditions. Consequently, a wide variety of storage equipment is commercially available to fulfil the enormous range of processing requirements of the various industries.

The container can have any shape that is practical (Figure 2.1), though for many applications they are cylindrical or rectangular with either flat (bin) or converging (bin-hopper) bases; grain height is often maintained at a value greater than two bin diameters (silo). Likewise, the shape, number and position of the discharge outlet can vary depending upon silo geometry, bulk solids properties and process requirements.

The geometry of the storage vessel and the characteristics of the stored material, particularly its frictional properties, ultimately dictate the flow patterns and the resulting pressure profiles developed within the bulk solids during filling and discharge. The two basic flow patterns that most commonly occur in silos are illustrated in Figure 2.2. In a 'mass-flow' type of hopper, all the contents of the vessel move during outloading, while in a 'funnel-flow' type, solids movement is confined to a vertical region in the centre of the vessel. Combination of the above two basic flow patterns is

Pressure profiles in bulk solids storage vessels 21

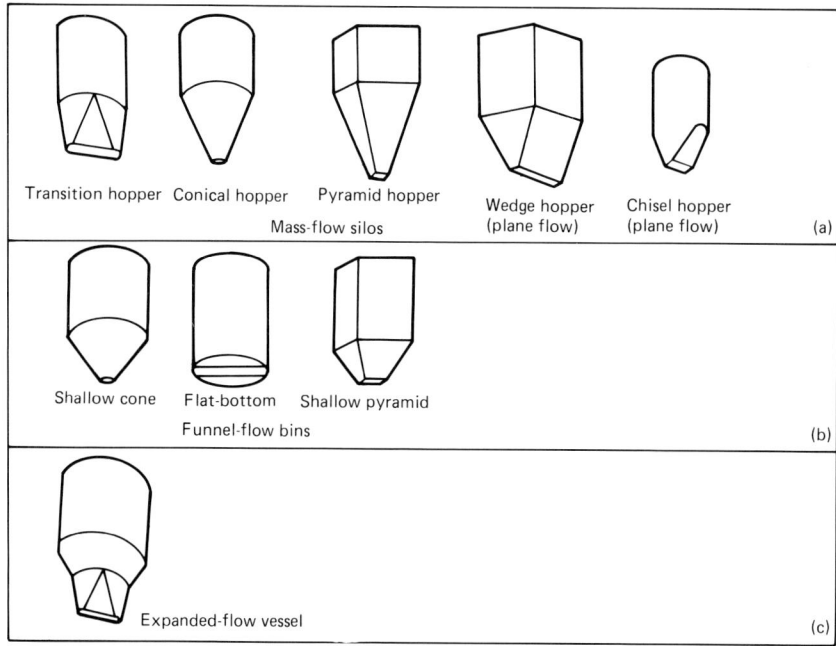

Figure 2.1 Typical vessel geometries for the storage of particulate.

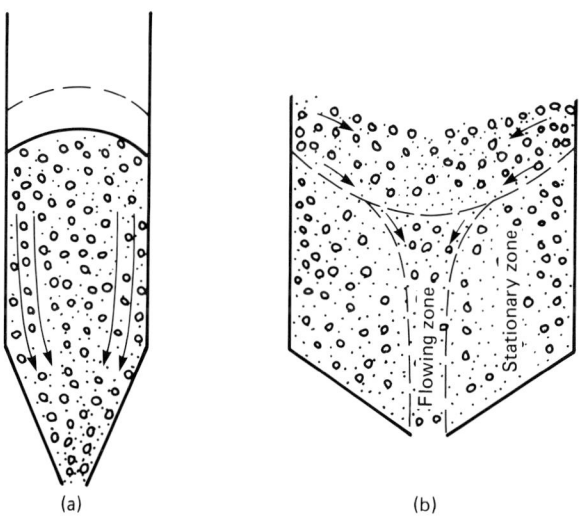

Figure 2.2 Flow patterns in (a) mass-flow (b) funnel-flow vessels.

also possible as in the 'expanded-flow' design shown in Figure 2.1c. Table 2.1 lists some of the advantages and disadvantages of each type.

The stresses within the bulk solids and at the boundary between the moving grains and the walls of the container are not isotropic and do not increase hydrostatically; the pressure increases almost exponentially with grain height to an asymptotic value, which in most cases, is reached at a critical depth of about 2–5 bin diameters. This is seen in Figure 2.3(a), which shows the vertical and lateral stress profiles for both the vertical and the converging sections of a bin–hopper system under the initial filling and static storage conditions. Figure 2.3(a) also illustrates the prevailing static stress field which is characterized by major vertical (consolidating) stresses that are parallel with the near-vertical lines of major principal stress.

It is also a well-established experimental fact that solids discharge causes substantially higher wall pressures than those predicted by the Janssen filling formula; overpressure factors of the order of 2.0 (Ravenet, 1981) to as high as 13 (Sundram and Cowin, 1979) have been reported. This dynamic overpressure during outloading is attributed mainly to a 'switch' in the stress field within the silo from the static state which exists during filling to a dynamic field during flow; the latter state of stress is characterized by the near horizontal lines of major principal stress. This is seen in Figure 2.3(b), which shows the variation of both lateral and vertical pressures with grain height during discharge.

The situation is further complicated because the changeover from static to dynamic stress field is not instantaneous, but transient in nature; the switch commences at the orifice as soon as flow is initiated. This is

Table 2.1 Advantages and disadvantages of mass-flow and funnel-flow silos

Mass-flow	Funnel-flow
Advantages	
1. Uniform and controlled flow with a constant bulk density	1. Can be made to fit available space
2. No arching, piping, channelling, surging or flooding	2. Fairly cheap with long life as there is no wear on the walls
3. First-in, first out; no dead zones; thus suited to materials which might deteriorate with time	3. Can be used to store more than one product during its life-time. Thus, economical
4. No segregation during storage or discharge	
5. Flow commences easily and as soon as the outlet is opened. The effect of time consolidation may be accounted for at design stage	
Disadvantages	
1. Usually tall with steep hopper, thus requiring a fair amount of head room	1. Uneven and unpredictable flow
2. Short life as there is material slip at the walls	2. Arching, piping and flooding problems are common
3. Specific to one material or group of materials. Thus, expensive	3. First-in, last-out
	4. Segregation will occur if the stored product has particle size distribution
	5. Dead pockets exist. Thus, not suited to perishable products

manifested by the presence of a large overpressure wave which, once established, is propagated rapidly upwards through the material.

In the case of most mass-flow silos, the propagation terminates at the point of intersection of the hopper and the vertical section (Figure 2.3b). With the switch permanently fixed at the point of transition, the overpressure is maintained for as long as the flow continues. Furthermore, with the overpressure effectively locked at the transition, the static stress field in the upper vertical section of the silo remains unaffected during discharge. This has been verified experimentally by Jenike et al. (1973) who observed that any increases on filling pressures in the vertical section during flow are primarily due to surface imperfections on the walls of the container, e.g. welded joints that have not been properly butted; such surface irregularities act as centres at which the switch in pressure field may occur. In practice, however, the resulting pressure fluctuations are unstable and, as a consequence are difficult to predict with confidence.

With flat-bottom bins and silos having shallow hoppers, discharge is more likely to be funnel-flow (Figure 2.2). In such circumstances, it is

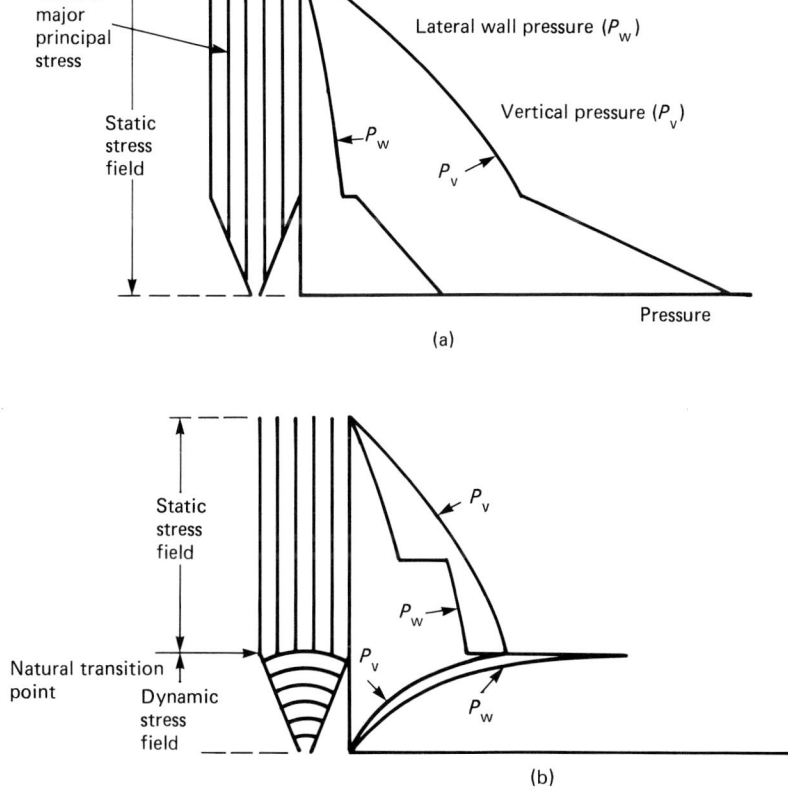

Figure 2.3 Pressure profiels in mass-flow silos. (a) Initial pressure profiles; (b) Discharge pressure profiles.

24 Pressure profiles in bulk solids storage vessels

difficult to ascertain the exact location of the effective transition at which the switch from static to dynamic stress field occurs. With deep bins, the central moving channel diverges upwards from the outlet and intersects the walls of the container at an effective transition point (Figure 2.4). The position of the effective transition depends largely on the half angle θ that the active channel makes with the vertical. The angle θ, in turn, depends upon the frictional properties of the material; in general, the effective transition point moves upwards towards the level of the free surface as the bulk material becomes less free flowing. Above the point of effective transition, solids movement is as in mass-flow. Below this point, the stagnant material between the live channel and the walls of the container acts as a buffer, dampening out the pressures that might otherwise be transmitted to the walls. Such dead zones are undesirable particularly when perishable and/or segregating products are stored.

In practice, the primary aim in silo design is therefore to promote mass flow, which means that the design is carried out with a specific material or group of materials in mind, since mass-flow for one particulate may be, and often is, funnel-flow for another. However, for economic reasons, it is quite common to use the same silo to store more than one product during its lifetime, simply because a well-designed silo with a reasonable degree of structural integrity often lasts considerably longer than the material for which it is initially designed. The alteration in material bulk and frictional properties that accompany such a change in product specification could have a detrimental effect on the flow patterns, often changing a mass-flow silo to a funnel-flow type. Moreover, for some materials it is extremely difficult to create mass-flow no matter how steep the walls or the size of the discharge. As a result, numerous funnel-flow silos are in existence today and their numbers are likely to go up even further in future.

Knowledge of the pressures acting in the fill and at the walls of the silo is essential not only for the structural design, but also for determining the critical silo dimensions in order to ensure unobstructed discharge; failure

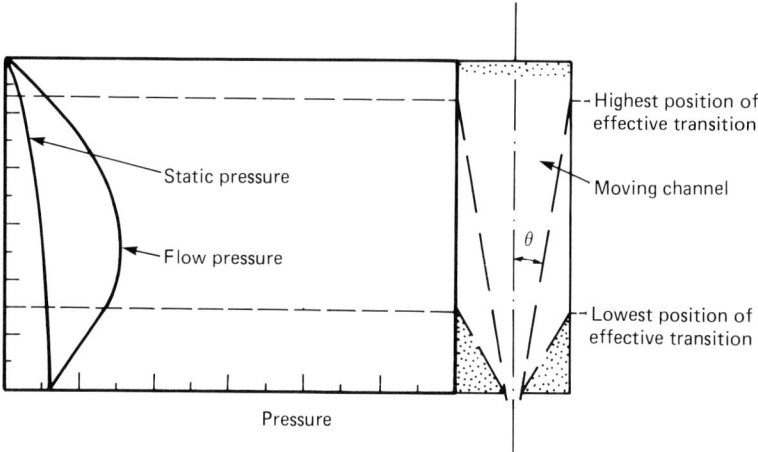

Figure 2.4 Pressure profiles in funnel-flow bins.

Pressure profiles in bulk solids storage vessels 25

to account for these pressures will result in unsatisfactory installations that are subsequently expensive and time-consuming to put right. The cost of correcting flow problems in processes that handle solids are substantially higher than those using fluids. Moreover, reports of both catastrophic collapses (Figure 2.5) and, less spectacular but equally serious, cracked walls appear regularly in the open literature (Jenike and Johanson, 1986). In the majority of cases, such failures are directly attributed to a lack of appreciation of the prevailing stresses in the silo.

Thus, in this chapter attention is focused upon pressure distributions in the silo during the charge, storage and gravity discharge; both mass-flow and funnel-flow silos are considered for centric filling and discharge. The analysis is based on the Janssen differential slice method and its extension by Walker (1966) and Walters (1973). This approach has been shown to be consistent with experimental observation and is easy to apply compared to more rigorous techniques such as the method of characteristics proposed by Sokolovski (1965).

Mechanical and structural strength of the silo to carry the prevailing stresses are not considered here. For information on this topic, the reader is refered to the publication of Reimbert (1975/76).

Variation of pressure with position in a particulate under storage

Initial filling and static pressures

(i) Vertical section (bin) — Janssen formula

For flat-bottom bins and the vertical part of deep silos, most investigators have correlated their pressure measurements during the initial filling and

Figure 2.5 Grain silo after collapse due to insufficient structural integrity.

26 Pressure profiles in bulk solids storage vessels

under storage using the classical Janssen equation developed nearly 100 years ago (Janssen, 1895). Despite the fact that most of the assumptions used in its derivation have been subsequently shown to be incorrect, the approach which is presented below is widely adopted in industry and is the basis of several pertinent design codes, e.g. German DIN 1055 (1984), American ACI Committee (1977) and Soviet CH-302-65 (Lipnitskii and Abramovitsch, 1967).

Consider the equilibrium of forces acting on an elemental slice of the stored powder of thickness, dh, at a depth h from the top surface in a deep bin with an overall height H, cross-sectional area A and circumference C as illustrated in Figure 2.6. The forces acting on the elemental slice are those on its two faces, sides and gravity. Let the vertical stress at the upper face, depth h, be P_v and that at the lower face, depth $h + dh$, be $P_v + dP_v$. P_w and S_w are the normal (horizontal) and tangential (shear) stresses at the wall respectively due to friction between the grains and the walls of the vessel. The weight of the powder within the slice is $Adh\rho_b g$ where g is the acceleration due to gravity and ρ_b represents material bulk density which is assumed to remain constant over the entire depth of powder. At the point of incipient motion, the algebraic sum of the forces in any direction must be zero. Resolving in the vertical direction:

$$P_v A + Adh\rho_b g - A(P_v + dP_v) + S_w C dh = 0 \tag{2.1}$$

To solve Equation 2.1, it is often assumed that for granular materials the ratio of horizontal to vertical pressure is constant everywhere in the element and is independent of the magnitude of the prevailing stresses. Thus:

$$P_w/P_v = K \tag{2.2}$$

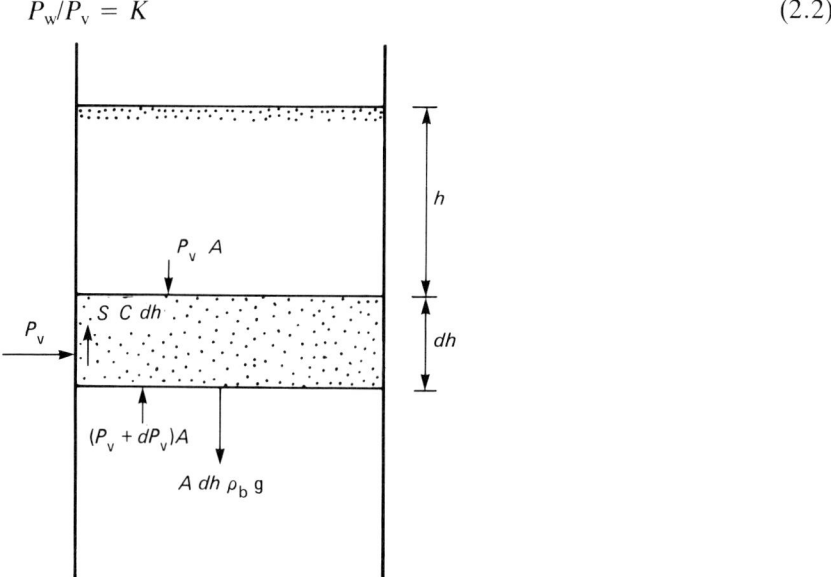

Figure 2.6 Forces acting on differential slice of fill in bin.

Moreover, with wall friction fully mobilized;
$$\mu = \tan \varphi_w = S_w/P_w \tag{2.3}$$
Expanding Equation 2.1, substituting for S_w using Equations 2.2 and 2.3 and simplifying yields the following first order differential equation:
$$dP_v/dh = \rho_b g - (\mu KC/A)P_v \tag{2.4}$$
Separating variables and integrating Equation 2.4 with the boundary condition $P_v = 0$ at $h = 0$, i.e. at the level of the free surface, and $P_v = P_v$ at $h = h$ gives:
$$P_v = \rho_b g A/\mu KC [1 - e^{-(k\mu C/A)h}] \tag{2.5}$$
and
$$P_w = P_v K = \rho_b g A/\mu C [1 - e^{-(k\mu C/A)h}] \tag{2.6}$$
for $h \to \infty$, i.e. for large depth of fill, the term in { } will be small and the expression in [ ] approaches unity. This gives the maximum asymptotic value of the vertical and normal wall stresses. These are:
$$P_v(\text{max}) = \rho_b g A/\mu KC \tag{2.7a}$$
and
$$P_w(\text{max}) = KP_v(\text{max}) = \rho_b g A/\mu C \tag{2.7b}$$
While for $h \to 0$, i.e. near the top surface, the exponential term in Equation 2.5 and 2.6 may be reduced to Equation 2.8 without significant loss of accuracy:
$$\exp - \{K\mu C/A\}h = [1 - (\mu KC/A)h] \tag{2.8}$$
Substituting in Equation 2.5 gives the expected vertical load near the top surface.
$$P_v(\text{top}) \simeq \rho_b g h \tag{2.9}$$
which is the hydrostatic pressure.

In the case of cylindrical silos, $A/C = D/4$, but in general this ratio is the hydraulic radius defined as:

A/C = Cross-sectional area/Perimeter

i.e.
$$A/C = D/2(1 + i) \tag{2.10}$$

Table 2.2 Hydraulic radius for common silo geometries

Silo shape	A	C	A/C
○	D/4	D	D/4
□	D	4D	D/4
▯ (plane-flow)	DL	2L (neglecting the ends)	D/2

where D = diameter or width

i = 1 for axisymmetric silos, e.g. circular or square

= 0 for plane flow bins, e.g. rectangular.

Table 2.2 gives values of the hydraulic radius for most commonly encountered silo geometries.

Thus, provided accurate material properties are available and that a reasonable estimate of the pressure coefficient, K, can be made, then vertical and horizontal (wall) Janssen stress distributions may be predicted using Equations 2.5 and 2.6.

Equations 2.5 and 2.6 are plotted in Figures 2.7–2.10 for circular silos. The curves indicate that the pressure at the wall and within the fill increases exponentially with depth until an asymptotic value is reached; the value of this maximum pressure is directly proportional to silo diameter and inversely proportional to the coefficient of wall friction (Equations 2.7 and 2.8). Below the critical depth, the stresses are independent of grain height. Consequently, the load on the bottom of the container is only a small fraction of the total weight of the material, most of the weight being supported by the side walls. In other words, the hydrostatic head of the stored material above the orifice point is not entirely available for flow, thus confirming the experimental observation that for deep bins the rate of solids discharge is practically unaffected by grain height.

However, extensive research over the past two decades has shown Janssen's original assumptions of constant coefficient of pressure, angle of wall friction and bulk density to be incorrect. The effect of variation in these parameters upon pressure profiles within the fill and at the walls of the container is discussed below.

The pressure coefficient, K, has a marked influence upon the pressure profiles within the powder and at the walls (Figure 2.10). Its value, however, is a matter of contention between investigators (Jenike et al., 1973, Harr, 1977 and Keimbert, 1976), for example according to Jenike et al. (1973) it has a universal value of 0.4 for most granular materials while others claim that K is a function of material angle of internal friction. Reimbert (1976) claims that the expression commonly used for K, with little theoretical justification (Sundram and Cowin, 1979), that is:

$$K = (1 - \sin \varphi)/(1 + \sin \varphi) = \tan(\pi/4 - \varphi/2) \tag{2.11}$$

is only true for an infinite depth of fill, and that the actual value is higher at small depths. Consequently, the mean vertical pressure in the upper region of the fill is less than that predicted by Equation 2.5 while the observed lateral pressure is higher than the estimated value from Equation 2.6. Reimbert allowed for this difference and suggested the following alternative equation for the lateral pressure in the case of circular bins:

$$P_w = \rho_b g D/4\mu [1 - \{h/((D/4\mu K) - (h_c/3)) + 1\}^{-2}] \tag{2.12}$$

with

$h_c = D/2 \tan \gamma$

and K given by Equation 2.11.

Pressure profiles in bulk solids storage vessels 29

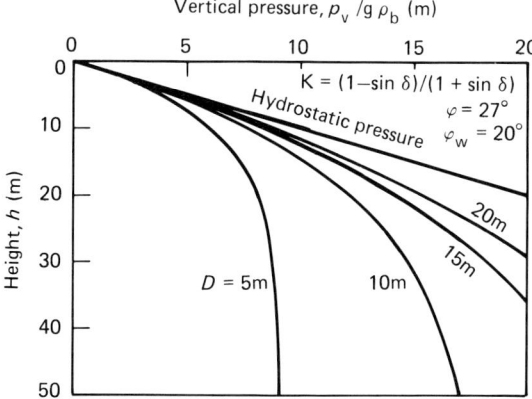

Figure 2.7 Influence of bin diameter on Janseen fill pressures.

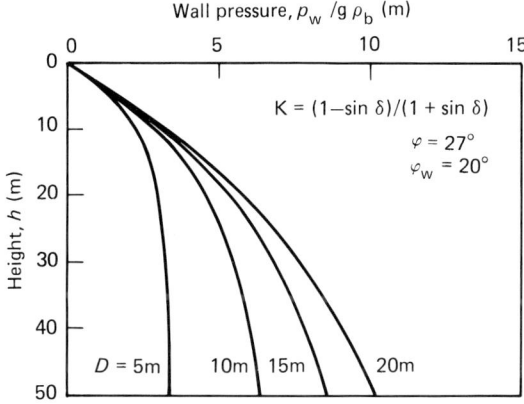

Figure 2.8 Influence of bin diameter on Janssen wall pressures.

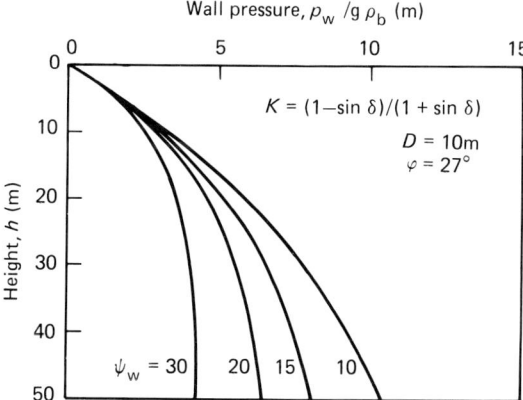

Figure 2.9 Influence of wall friction on Janssen wall pressures.

30 Pressure profiles in bulk solids storage vessels

For large depths of fill, Equation 2.12 coincides closely with the Janssen pressure formula (Equation 2.6), but deviates from it for small depths. Moreover, Equation 2.12 is only applicable to cylindrical bins, because Reimbert's experimental observation suggests that K also varies strongly with the shape of the silo (Figure 2.11).

The dramatic influence of variation in the angle of wall friction upon wall stresses may be seen from the curves in Figure 2.9. The two coefficients of wall and internal friction usually vary with voidage, temperature, moisture, storage time, and method of determination, (Schwedes, 1985). Furthermore, with the degree of consolidation, and thus voidage, varying with depth of fill, it is reasonable to expect the two coefficients of friction also to vary with depth. As a result, reported experimental measurements of the two angles of friction for apparently the same material often vary widely among investigators, making it difficult for the process engineer and equipment designer to decide on the correct value to be used with Equations 2.5 and 2.6.

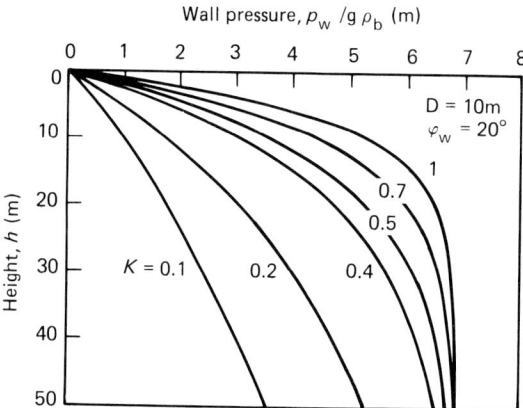

Figure 2.10 Influence of pressure coefficient on Janssen wall pressures.

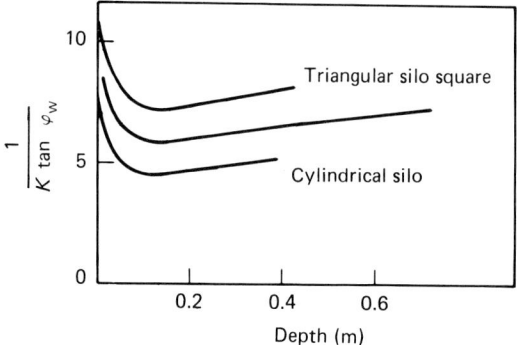

Figure 2.11 Variation of pressure coefficient with depth for different shapes but of same mean radius (Reimbert, 1975/6).

The apparent bulk density of most bulk granular materials also depends upon the degree of consolidation and, thus, depth of fill in the silo; increase in bulk density can vary widely from 20% for granular solids to as much as 80% for powders (Junior, 1983).

However, despite all these shortcomings Janssen's formula describes most experimental data with a fair degree of accuracy, (Bishara, 1983). Numerous other theories and solutions of varying complexity are available in the open literature (Enstad, 1975; Hancock and Nedderman, 1974; and Nedderman, 1981, Horne and Nedderman, 1976, 1978a, b) but on close examination they differ little from that of Janssen. Such differences are even less significant considering the uncertainty associated with the estimation of some of the bulk and frictional characteristics of the material and their variation in the silo. It is therefore reasonable to adopt Janssen's equation to predict the stresses during the initial filling and under static storage.

(ii) Converging section (hopper)

Janssen's hypothesis as developed above is only applicable to the upper section of deep silos and flat-bottom bins, ignoring any end-effects. For the hopper region, both Walker (1966) and Jenike et al. (1973) assumed a linear hydrostatic pressure gradient and proposed the following equation to describe the vertical pressure distribution:

$$P_v = \rho_b g[\{P_{vo}/\rho_b g\} + h] \qquad (2.13)$$

where P_{vo} is the vertical surcharge pressure at the top of the hopper (Figure 2.12) obtained from Janssen's relationship (Equation 2.5); the expression in { } is the surcharge head at this point. To determine the lateral wall pressure, both workers assumed that the ratio of normal wall pressure to vertical pressure is constant and given by Equation 2.2 with

$$K = \tan \alpha/(\tan \varphi_w + \tan \alpha) \qquad (2.14)$$

for Jenike and

$$K = \sin(2\alpha \cos t\, \varphi_w)/[\sin(\varphi_w + 2\alpha) + \sin \varphi_w] \qquad (2.15)$$

for Walker.

Equations 2.13–2.15 will be examined against experimental data later on in this chapter.

Dynamic discharge pressures

Mass-flow silos

The available experimental evidence indicates unequivocally that Janssen equations developed for the initial filling process underestimate the pressures exerted on the walls of the container during draw-down; experimental data suggest that factors of the order of 2 for granular materials and as much as 13 for powders are required in the Janssen formula in order to account for discharge pressures (Deutsch and Schmidt, 1969, Sundram and Cowin, 1979 and Revenet, 1981).

32 Pressure profiles in bulk solids storage vessels

Most investigators and design codes recognize the overpressure generated during the emptying process, but wide differences exist in their recommendations to allow for the dynamic effects. In most cases, the modifications are carried out either by the addition of an empirical dynamic factor to the basic Janssen formula, or by the alteration of any one or combination of the pertinent parameters in Equations 2.5 and 2.6. The approach is well illustrated by briefly examining parts of the new German DIN 1055 (1984), which is one of the most prominent design codes currently available. The new design is based on the Janssen formula, but unlike its predecessor it distinguishes between pressure patterns developed within the stored material during filling and emptying. Despite this recognition, its recommendation is that the value of the pressure coefficient, K, should be taken empirically as 0.5 and 1.0 for filling and discharging respectively. The code also confirms that the coefficient of wall friction, μ, is not uniquely fixed, and proposes three different values for any particular material ranging from extremely rough, e.g. silos of corrugated steel, to extremely smooth, e.g. metals and plastics. Finally,

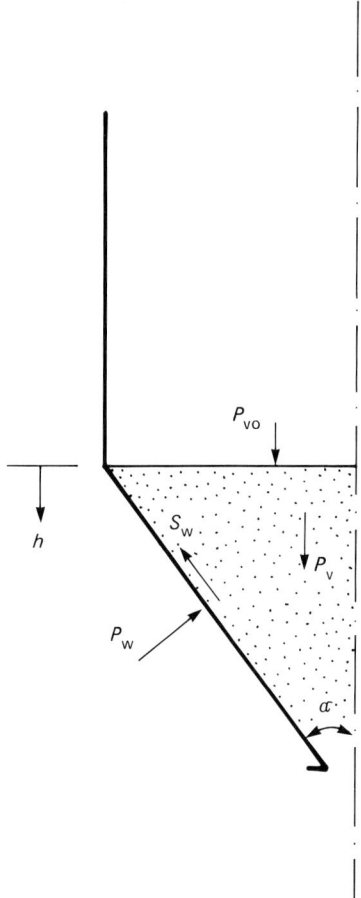

Figure 2.12 Forces acting in hopper.

with regard to the variation in material bulk density, the code recommends different values for the upper and lower sections of the silo.

In practice, such simple empirical modifications to Janssen pressure profiles are a gross oversimplification of the problem and do not account for some of the complex phenomena that occur during discharge, e.g. the presence of large local overpressure peaks often observed both with model and industrial silos. Indeed, most silo failures, both catastrophic collapse and less spectacular but equally serious cracked walls, are attributed directly to a lack of appreciation of these local overpressure peaks generated at the walls of the silo during grain discharge.

Solids discharge is accompanied by a dramatic alteration in the stress field from a static state under filling and storage conditions to a dynamic state during flow (Figure 2.3). The switch in the stress field which is most pronounced for the converging section, in turn, causes most of the load to be transferred to the walls of the container.

The switch and the associated overpressure wave is initiated at the outlet when discharge commences, and travels rapidly upwards through the bulk solids and finally becomes locked at an effective transition point which, in the case of most mass-flow silos, is the point of intersection of hopper and vertical walls (Figure 2.3b); at this point, it remains as a permanent overpressure load until flow stops.

(i) Vertical section (bin)

Above the point of transition, that is in the vertical part of the silo, the static stress field is unaffected, provided that the side walls are absolutely straight and free from surface irregularities (Jenike *et al.*, 1973). Under such circumstances, the pressure profiles in the vertical section may be predicted using Janssen's filling formula (Van Zanten and Mooij, 1977; Everts *et al.*, 1977 and Richards, 1977).

Jenike *et al.* (1973) point out that in commercial silos, the presence of local surface imperfections is inevitable. During mass flow, these local convergences control the formation of wall boundary layers, and thus determine the nature of the prevailing stress field. For perfectly straight walls with no surface irregularities, wall boundary layers are absent and the initial Janssen stress field prevails. Where boundary layers exist flow pressures are predicted by assuming that the elastic strain energy within the flowing mass tends towards a minimum.

The change between the stress fields initiated by local convergences in the cylinder occurs at a switch at some depth, h, in the channel. Therefore the initial Janssen pressure field applies between the top of the container and the switch; below this level, that is between the switch and the level of the free boundary condition, the minimum strain energy applies. However, due to the random nature of such surface irregularities, it is difficult to predict the exact location of the switch and the consequential overpressure peak in advance. Moreover, Jenike *et al.* (1973) point out that such a switch in the stress field is unstable and reverts back to its original form some distance away from the local disturbance causing it. Thus, for deep silos, it is likely that there will be more than one switch during flow, each with an overpressure peak.

To overcome this difficulty, Jenike et al. (1973) suggest the evaluation of a bound enclosing all possible pressure peaks using their minimum strain–energy stress field theory. This is principally based on the second law of thermodynamics; it states that during steady discharge of solids, energy is expended at the maximum possible rate and, as a result, the internal energy of the system will tend to a minimum. For relatively large silos, the kinetic energy terms may be ignored, and provided no temperature and chemical changes occur during the handling operation, the only component of the internal energy that will tend to minimize is the elastic strain energy. Jenike et al. (1973) proposed that the recoverable and the non-recoverable strain energy terms in the governing equations describing the flowing solids will determine whether a switch in the stress field will occur or not. They thus formulated the problem by writing an energy balance for the system, and solved it for the case of vertical cylinders using variational calculus to minimize the recoverable strain energy. Bounds for all possible pressure peaks were provided in this way by Jenike et al. (1973), but in general the determination of such bounds involves enormous computational effort.

McLean and Arnold (1976) examined the instantaneous pressure distribution curves provided by Jenike's strain–energy theory and noticed that, in each case, the pressure peak occurs at the level of the switch itself and that the magnitude of the peak stress is greatest when the natural boundary condition is taken to be the transition between the cylinder and the hopper. These workers also noted that for most commercial materials and silo geometries, the location of the switch is in the range $0.5H$ to $0.7H$ for axisymmetric silos and $0.4H$ to $0.65H$ for plane-flow vessels. They thus suggested that in most situations, a single calculation is sufficient to provide a reasonable estimate of the maximum peak strain–energy pressure. They proposed that for the purpose of this evaluation, the location of the switch should be taken at $0.6H$ and $0.5H$ for axisymmetric and plane-flow geometries respectively. Their modified equation for the single maximum pressure peak is given by:

$$P_w(\max) = A\rho_b g/C \tan \varphi_w [1 - (R - Q) \tan \varphi_w / M^i] \tag{2.16}$$

with

$$R = \frac{-(K_h M^i - 1)(S_0 - N) e^{-x} + M^i(\mu^{-1} - K_h N)}{(k_h M^i + 1) e^x - (K_h M^i - 1) e^{-x}} \tag{2.17}$$

where

$$K_h = v/1 - v \tag{2.18}$$

$$S_0 = 1/\mu K(1 - \exp - [\mu K h_s C/A]) \tag{2.19}$$

$$R + Q = S_0 - N$$

$$N = 2v/\mu M^{2(i-1)} \tag{2.21}$$

$$M = \sqrt{2(1-v)} \tag{2.22}$$

$$v = K/(1 + K) \text{ for axisymmetric flow} \tag{2.23a}$$

$$= \frac{K + 2 - \sqrt{4 - 3K^2}}{2(K + 1)} \text{ for plane flow} \tag{2.23b}$$

$$Z = H - h_s/M^i(A/C) \tag{2.24}$$

with

$h_s = 0.6H$ for axisymmetric flow
   $= 0.5H$ for plane flow

Below the level of effective transition, the pressure is obtained on the basis of 1.5 times Janssen's filling formula, but for design purposes the maximum pressure at the level of the effective transition is used down to the top of the hopper. Above the level of transition, up to the free surface at the top, Janssen's filling formula may be adopted to estimate the prevailing pressure together with Equation 2.11 for the pressure coefficient, K.

(ii) Converging section (hopper) with surcharge

Walker (1966) and Walters (1973) extended the elemental slice method of Janssen in order to analyze stresses in the converging section of the hopper. The analysis that follows is based on the differential slice method and is primarily due to these workers.

Consider the vertical equilibrium of a horizontal slice of material of thickness dh in a conical hopper (Figure 2.13). The stresses acting on the

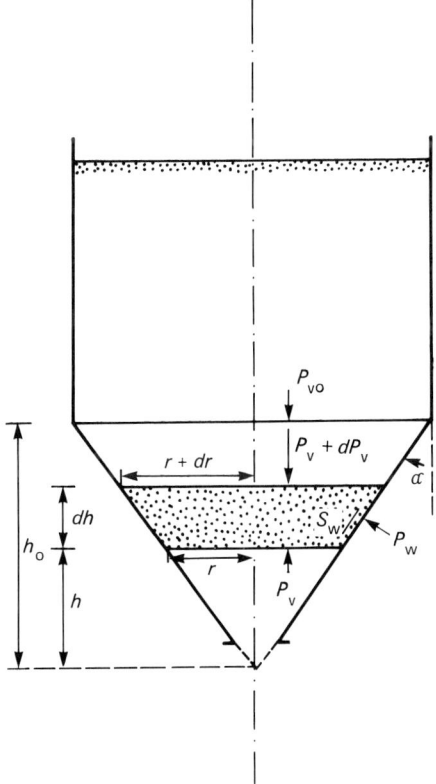

Figure 2.13 Forces acting on differential slice of fill in hopper.

elemental slice at depth h and $h + dh$ above the apex are P_v and $P_v + dP_v$ respectively, the weight of the slice $\rho_b g\, dh\, \pi r^2$, and P_w and S_w, the normal and shear stresses at the wall. Assuming mass flow and constant vertical stress, P_v, over the elemental slice, the vertical equilibrium of forces may be written as:

$$(P_v + dP_v)\pi(r + dr)^2 + \pi r^2 dh \rho_b g = P_v \pi r^2$$
$$+ P_w \sin \alpha \{2\pi r(dh/\cos \alpha)\} + S_w \cos \alpha \{2\pi r(dh/\cos \alpha)\} \qquad (2.25)$$

where the expression in { } is the area of contact of the powder element at the wall.

Expanding Equation 2.25, ignoring higher-order terms such as dr^2 and $drdP_v$, putting $r = h \tan \alpha$ and rearranging, yields

$$2P_v dh/h + dP_v + \rho_b g dh = (2dh/h \tan \alpha)P_w(\tan \alpha + \tan \varphi_w) \qquad (2.26)$$

The relation between the vertical stress, P_v, and the normal stress at the wall, P_w, is obtained from the Mohr circle representing the stress conditions at the wall (Figure 2.14). The major principal stress is in the direction shown by P_1 and the major principal plane is at right angles to it. β is the angle between the wall and the major principal plane and α is the hopper half-angle.

Assuming the switch from static to dynamic stress field occurs as soon as discharge is initiated, and that the material yields within itself as it passes through the converging channel, the stresses in the powder close to the wall are represented by a Mohr circle which just touches the powder yield locus OM. Further, as mass flow prevails, the material must be slipping at the wall and therefore the wall stresses are represented by a point on the wall yield locus ON. In the present case of a converging channel, this is given by point e (and not e'). Thus, OE represents the magnitude of the normal wall stress P_w.

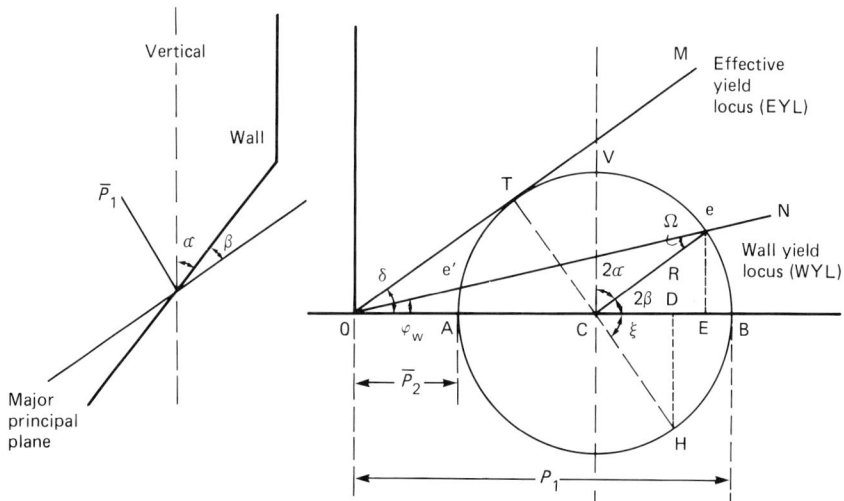

Figure 2.14 Stress conditions at hopper wall.

On the Mohr circle, points H and V represent the horizontal and vertical directions respectively. The magnitude of the vertical stress, P_v, is represented by OD.

From the geometry of the Mohr circle:

$$OD = OC + CD = R/\sin \delta + R \cos \xi \qquad (2.27a)$$

$$= R/\sin \delta - R \cos(2\beta + 2\alpha) \qquad (2.27b)$$

while

$$OE = OC + CE = R/\sin \delta + R \cos 2\beta \qquad (2.28)$$

therefore

$$P_w/P_v = OE/OD = (1 + \sin \delta \cos 2\beta)/[1 - \sin \delta \cos(2\alpha + 2\beta)]$$

$$= K_1 \qquad (2.29)$$

β may be expressed in terms of φ_w and δ by referring to the geometry of the Mohr circle and noting that

$$2\beta = \varphi_w + \Omega \qquad (2.30)$$

Considering the triangle OCT and OCE and using the sine rule:

$$\sin \Omega / \sin \varphi_w = OC/R = \sin \delta \qquad (2.31)$$

Thus,

$$2\beta = \varphi_w + \sin^{-1}(\sin \varphi_w / \sin \delta) \qquad (2.32)$$

Replacing P_w in Equation 2.26 using Equation 2.29 and simplifying gives the following first order differential equation:

$$dP_v/dh = K_2 P_v/h - \rho_b g \qquad (2.33)$$

where

$$K_2 = 2(K_3 - 1) \qquad (2.33a)$$

$$K_3 = K_1[(\tan \alpha + \tan \varphi_w)/\tan \alpha] \qquad (2.33b)$$

Equation 2.33 is a homogeneous first-order differential equation which can be solved by using an integrating factor. The final solution for P_v is:

$$P_v = \rho_b g h/(K_2 - 1)[1 - (h_0/h)^{K_2 - 1}] + P_{vo}(h/h_0)^{K_2} \qquad (2.34)$$

which describes the variation of the vertical stress with grain height in converging section of the silo. The distribution of the normal stress at the wall may be obtained from equations 2.34 and 2.29. Thus,

$$P_w = K_1\{\rho_b g h/(K_2 - 1)[1 - (h_0/h)^{K_2 - 1}] + P_{vo}(h/h_0)^{K_2}\} \qquad (2.35)$$

With no surcharge present, i.e. $P_{vo} = 0$, Equations 2.34 and 2.35 reduce to:

$$P_v = \rho_b g h/(K_2 - 1)[1 - (h_0/h)^{K_2 - 1}] \qquad (2.36)$$

and

$$P_w = K_1 \rho_b g h/(K_2 - 1)[1 - (h_0/h)^{K_2 - 1}] \qquad (2.37)$$

respectively.

38 Pressure profiles in bulk solids storage vessels

Walker (1966) and Walker and Blachard (1967) checked Equations 2.34 and 2.35 for flow of coal powders in a model silo with surcharge. Subsequent experimental data (Aoki and Tsunakawa, 1969) have shown that Equations 2.34 and 2.35 give results in the converging section with surcharge that are fairly close to measured values under dynamic conditions, but does not give agreement in the vertical section. For hoppers with no surcharge, Aoki and Tsunakawa (1969) reported experimental wall stresses significantly higher than values predicted by Equations 2.36 and 2.37; for any hopper geometry, the degree of deviation increases with increasing rate of discharge and position from the surface of the cone (Figures 2.15 and 2.16).

Experimental measurements of Aoki and Tsunakawa (1969) and Walker (1966) are used to test the validity of the various models presented in this chapter (Figures 2.17–2.21). The results of this investigation indicate that a

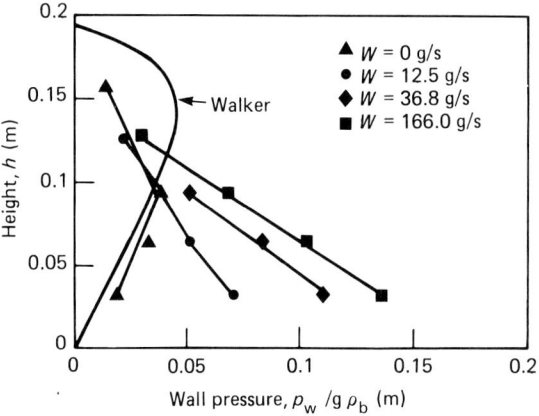

Figure 2.15 Experimental and theoretical hopper wall loads during discharge with no surcharge. Data from Aoki and Tsunakawa (1968): (α - 20°, φ_w = 19.9°, φ = 42.6°

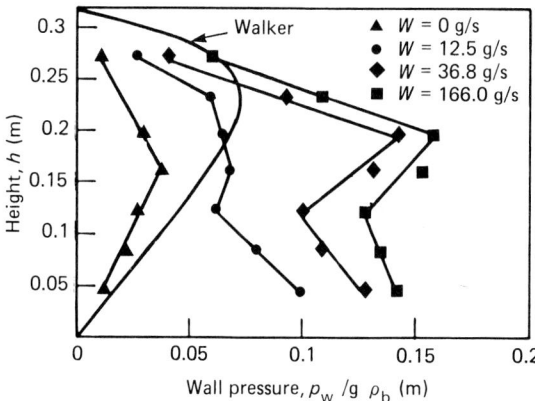

Figure 2.16 Experimental and theoretical hopper wall loads during discharge with no surcharge. Data from Aoki and Tsunakawa (1968): α = 30°, φ_w = 19.9°, φ = 42.6°.

complete assessment of the pressures within the fill and those exerted at the walls of a silo requires four separate analyses:

(a) In the vertical part of the silo, during the initial filling and under static storage conditions Janssen's Equations 2.5 and 2.6 are adequate with the static pressure coefficient given by Equation 2.11.

(b) In the vertical section of the silo, during discharge, prediction of wall pressures may be based upon Janssen's equations as in (a) provided the walls of the bin are perfectly straight and free from surface imperfections. This condition may be achieved with model bins, but not with industrial silos; Jenike's strain–energy theory and its modification by

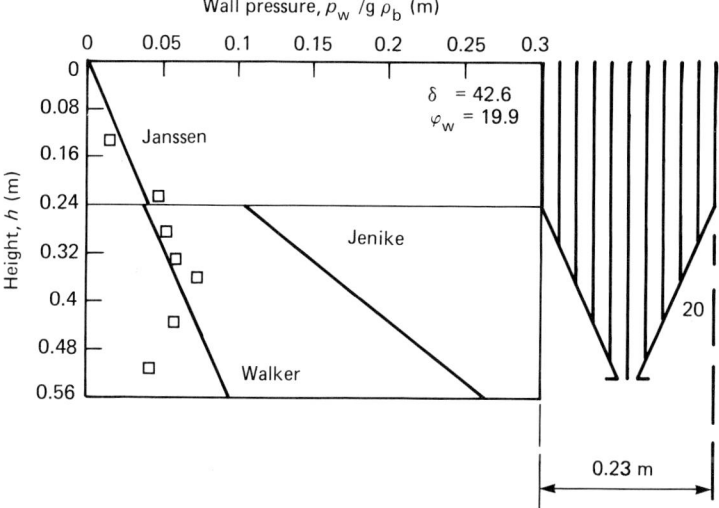

Figure 2.17 Experimental and theoretical wall pressure in silo during filling. Data from Aoki and Tsunakawa (1968)

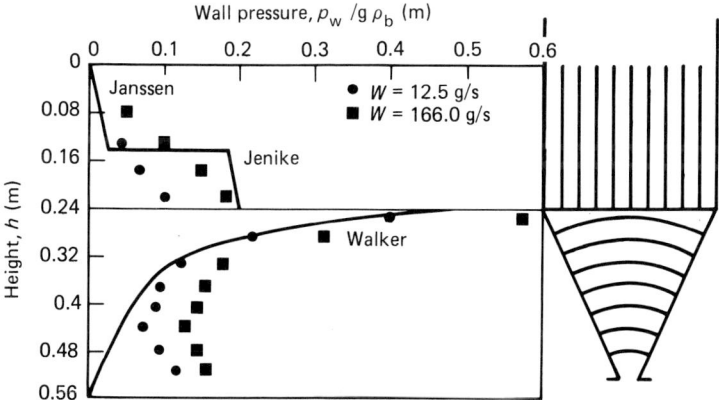

Figure 2.18 Experimental and theoretical wall pressure in silo during discharge. Data from Aoki and Tsunakawa (1968)

40   Pressure profiles in bulk solids storage vessels

McLean and Arnold (1976) cater quite well for industrial containers with slight surface irregularities (Equations 2.16—2.24).

(c) In the converging section during filling and under storage conditions either Walker/Walker's theory (Equations 2.13 and 2.15) or Jenike's approach (Equations 2.13 and 2.14) may be used; the latter tends to overestimate wall pressures (Figures 2.17 and 2.19).

(d) In the converging section during draw-down, Equations 2.34 and 2.35 give good agreement with experimental observations, provided the rate of discharge is not excessively high (Figures 2.18 and 2.20).

In the converging part of the silo, wall load increases with increasing rate of discharge (Figures 2.18 and 2.20); Griffith (1983) reported that a silo which had given 15 years of satisfactory service began to fail when the rate of draw-down was increased by 20%. None of the current theories can

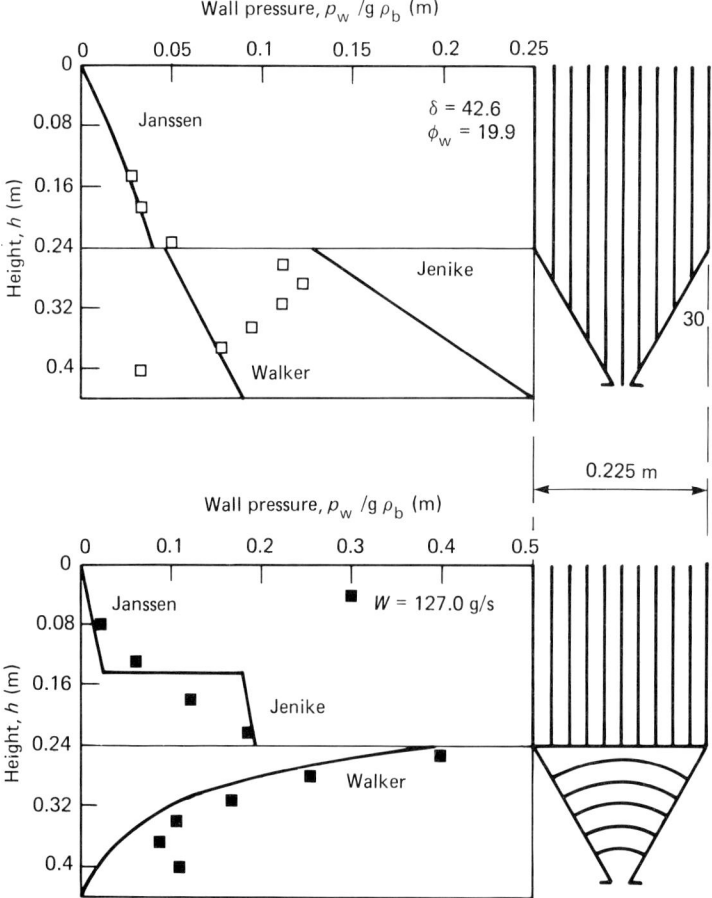

Figure 2.19 (top) Experimental and theoretical wall pressure in silo during filling. Data from Aoki and Tsunakawa (1968)

Figure 2.20 (above) Experimental and theoretical wall pressure in silo during filling. Data from Aoki and Tsunakawa (1968)

cater for the effect of variation in the rate of discharge upon wall pressures during flow.

In large silos with multiple outlet and inlet chutes, non-uniform pressure distributions may arise because of eccentric discharge and filling particularly when all draw-off points are not operated simultaneously (Rotter, 1983). Eccentric wall loadings may also arise because of off-wall discharging (Roberts and Ooms, 1983) or as a result of non-uniform air distribution near the outlet; asymmetrically located air injection points are used as an aid to silo discharge (Williams et al., 1983).

However, despite a number of investigations on eccentric flow (Jenike, 1967 and Roberts and Ooms, 1983), the complex nature of asymmetric flow is less amenable to theoretical analysis compared with centric loading and off loading. Consequently, prediction of the position of the switch and the resulting wall pressure is difficult to ascertain with precision. For single eccentric discharge, Roberts and Ooms (1983) recommended Ghowral's approach which involves the application of an overpressure factor, C_d, on Janssen's Equation 2.6. C_d is defined as:

$$C_d = K_a/K \tag{2.38}$$

$$K_a = \frac{1 + \sin \delta \cos 2\beta}{1 - \sin \delta \cos 2(\alpha + \beta)} \tag{2.39}$$

with β given by Equation 2.32. α is the 'pseudo' hopper half-angle based on the slope of the flow channel and K is the Janssen static coefficient of pressure with a value of 0.4.

A further shortcoming of the existing theories is that wall pressures are estimated assuming rigid silo walls. Roberts and Ooms (1983) point out that excessive cyclic variations in atmospheric temperature would cause cyclic expansion and contraction in the shell of the silo. This, in turn, increases wall pressures particularly when storing bulk materials over long periods of time without discharging.

Simultaneous charging and discharging also influences pressures prevailing in the silo (Wood, 1983; Sugden, 1980). In spite of this, little

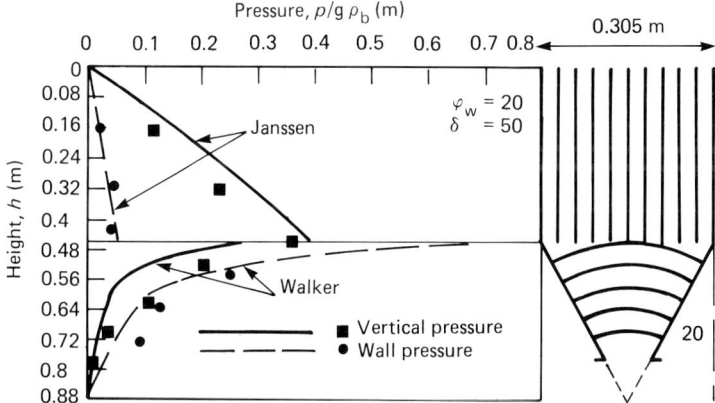

Figure 2.21 Experimental and theoretical wall and vertical fill pressures during steady discharge in silo. Data from Walker (1966).

42 Pressure profiles in bulk solids storage vessels

theoretical work has been reported in the open literature to account for the effect of concurrent filling and draw-down upon pressures within the fill and on the walls of the silo.

In summary, while existing theories cater well for gravity mass-flow hoppers with centric filling and discharge, prediction of wall loads during concurrent loading and draw down, for asymmetric flow and for high rates of discharge is less certain. In the absence of any fundamental work in cases where such complex flow problems are present, design is based on previous experience and large overpressure factors are employed.

Funnel-flow silos

So far, this chapter has concentrated on mass-flow silos (Figure 2.2). For flat-bottom bins and silos with shallow hoppers, discharge is more likely to be funnel-flow in which solids movement is confined to the central region of the container. The size and shape of the live core and the resulting pressure patterns depend critically upon the geometry of the silo and the frictional and bulk properties of the material (Figure 2.22), (Murfitt, 1980; Murfitt *et al.*, 1981).

The walls of the live core and the stagnant solids in its close proximity are inherently unstable, forming and collapsing continuously. Consequently, funnel-flow silos are more erratic in their behaviour and therefore less desirable than mass-flow hoppers. Moreover, in funnel-flow bins, the interface between the static and the dynamic zones does not occur at the natural transition of hopper and the vertical walls, but varies from a lower limit at a height approximately 1–2 bin diameters to an upper bound which could extend to the level of the free surface; the exact position of the upper bound of effective transition depends upon silo geometry, and material frictional characteristics (Figure 2.4). For the interface to occur at the level of the free surface, Hazra and Basur (1980) suggest:

$$H/D = K_4 = \tan \delta + \sqrt{\{\tan \delta[(1 + \tan^2 \delta)/(\tan \delta + \tan \varphi_w)]\}} \quad (2.40)$$

For shallow bins and silos, i.e. $H/D < 2$, (Figure 2.22a) the live channel rarely intersects the walls, and as a result no significant pressure peak occurs. In such circumstances, the Janssen formula may be adopted for design purposes.

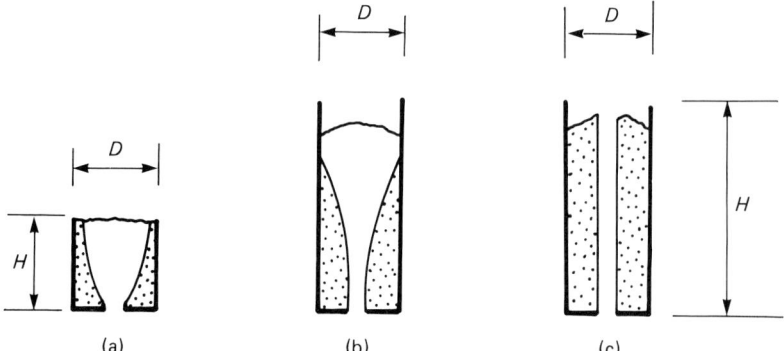

Figure 2.22 Typical flow patterns in funnel-flow bins.

For deep bins and silos with $H/D > 5$, the interface between the static and the flow zone rarely reaches the level of the free surface (Figure 2.22b). In such a case, the upper plug flow region may be designed on the bases of the Janssen formula or Jenike's strain–energy theory. Below the lowest position of effective transition, a linear pressure gradient may be assumed down to Janssen limiting value at the outlet.

The point of effective transition is usually expressed in terms of the half-angle θ between the live channel and the vertical (Figure 2.15). The value of θ in turn depends upon the effective angle of internal friction and wall friction; it varies, for a given powder, depending upon its moisture content, temperature, storage time, filling and discharge sequence. Hazra and Basur (1980) suggest that the highest position of the effective transition occurs when

$$\theta = \tan^{-1} 1/2(1/K_4) \tag{2.41}$$

while Arnold et al. (1980) recommends that θ should be taken as;

$$\theta = 40 \exp - [(\delta - 20)/(6 + \delta/6.5)] \tag{2.42}$$

Giunta (1969) provides a plot of θ as a function of the effective angle of internal friction (Figure 2.23).

Between the highest and the lowest position of the effective transition, a locus of pressure peak at the wall may be estimated from Jenike's modification of the Janssen pressure formula (Jenike et al., 1973):

$$P_w = K_5(\rho_b g A/\mu KC)[1 - e^{-(\mu KC/A)H_e}] \tag{2.43}$$

where H_e is the height of solids above the point of effective transition and the constant K_5 is given by Ooms and Robert (1985) as:

$$K = \frac{(24 \tan \theta + \pi/q)(1 - \sin \delta \tan \theta)}{16(\sin \delta + \tan \theta)} \tag{2.44}$$

with

$$q = \frac{\pi}{24 \sin \theta} \left\{ \frac{2Y}{(X - 1) \sin \theta} [\tan \theta + \sin \delta] - 1 \right\} \tag{2.45}$$

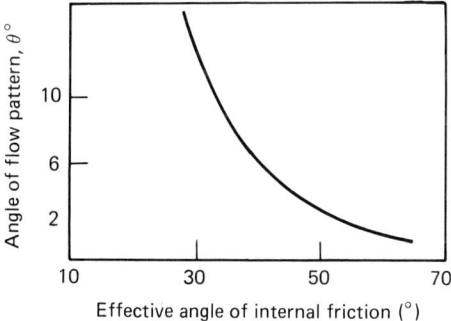

Figure 2.23 Variation of angle of flow with effective angle of internal friction in funnel-flow bins (Giunta, 1969).

44 Pressure profiles in bulk solids storage vessels

$$X = \frac{2^i \sin \delta}{1 - \sin \delta} \left[\frac{\sin (2\beta + \alpha)}{\sin \alpha} + 1 \right] \quad (2.46)$$

$$Y = \frac{[2(1 - \cos (\beta + \alpha))]^i (\beta + \alpha)^{1-i} \sin \alpha + \sin \beta \sin^{1+i}(\beta + \alpha)}{(1 - \sin \delta)\sin^{2+i}(\beta + \alpha)} \quad (2.47)$$

In Equation 2.47 the term $(\beta + \alpha)$ in the numerator is in radians and

$i = 0$ for plane flow geometry
$i = 1$ for axisymmetric flow

All other parameters have been defined previously.

More complex flow patterns have been reported for funnel-flow bins and silos by Giunta (1969) Kvapil (1965) and Deutsch and Schmidt (1969) and Van Zanten et al. (1977). These workers observed a number of distinct zones within the powder (Figure 2.24). The grains in the feed zone A slide over particles in zone B which, in turn, slide over the grains in the dead zone E. In the pipe or flow zone C, the grains move in arcs accelerating towards the free fall zone D in which the grains fall with a velocity approximately 1% of their terminal velocity. In the upper part of the bin, zone F, particles move in plug flow.

Figure 2.25 shows variation of wall pressure with grain depth for the geometry used by Van Zanten et al. (1977) in their study on funnel-flow bins. A large amount of scatter was observed in the lateral pressure during flow, particularly with silo walls that had surface imperfection (Figure 2.25). However, as seen, Equations 2.40–2.43 are consistent with the mean discharge pressures.

During the initial filling and under static conditions, the Janssen formula (Equations 2.5 and 2.6) may be used as in mass flow.

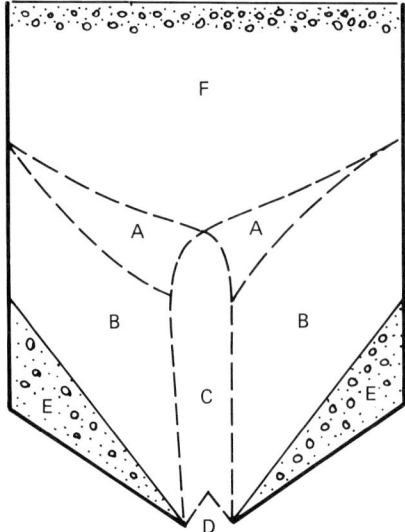

Figure 2.24 Flow pattern for gravity discharge in funnel-flow silo.

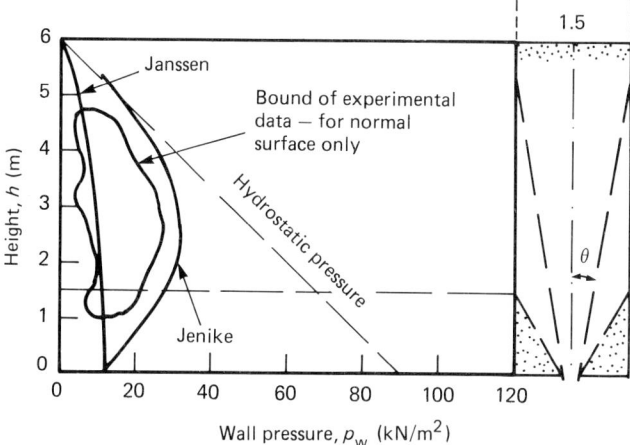

Figure 2.25 Experimental and theoretical wall pressure profile during discharge in a funnel-flow bin with a flat bottom. Data from Van Zanten et al. (1977). $\delta = 37.5°$, $\varphi_w = 23.5°$, $\rho_b = 1525$ kg/m^3, $K = 0.4$

References

ACI Standard 313–77 (1977, March). Americal Concrete Institute
AOKI, R. and TSUNAKAWA, H. (1969). *J. of Chem. Eng. of Japan*, **2**, 126–129
ARNOLD, P. C., MCLEAN, A. G. and ROBERTS, A. W. (1980). Bulk Solids Storage, Flow and Handling, TUNRA, The University of Newcastle, Australia
BISHARA, A. G. (1983). *2nd Int. Conf. on Design of Silos for Strength and Flow*. 49–65. Powder Advisory Centre, London
DEUTSCH, G. P. and SCHMIDT, L. C. (1969). Trans. ASME, *J. of Eng. for Ind.*, **91B**, (2), 450–459
DIN 1055, Part 6, Yellow print, 1984
ENSTAD, G. (1975). *Chem. Eng. Sci.*, **30**, 1273–1283
EVERTS, R., VAN ZANTEN, D. C. and RICHARDS, P. C. (1977). Trans ASME. *J. of Eng. for Ind.* **99**, (4), 4, 824–827
GIUNTA, J. C. (1969). Trans. ASME, *J. of Eng. for Ind.*, **91**, (2), 406–413
GRIFFITH, J. (1983). *2nd Int. Conf. on Design of Silos for Strength and Flow*. 14–23. UK Powder Advisory Centre, London
HANCOCK, A. W. and NEDDERMAN, R. M. (1974). *Trans. Inst. Chem. Eng.* **52**, 170–179
HARR, M. E. (1977). *Mechanics of Particulate Media*. McGraw-Hill Int.
HAZRA, S. K. and BASUR, H. S. (1980). *Proc. Int. Conf. on Design of Silos for Strength and Flow*, Univ. of Lancaster, UK Powder Advisory Centre, London
HORNE, R. M. and NEDDERMAN, R. M. (1976). *Powder Technol.* **14**, 93–102
HORNE, R. M. and NEDDERMAN, R. M. (1978a). *Powder Technol.* **19**, 235–241
HORNE, R. M. and NEDDERMAN, R. M. (1978b). *Powder Technol.* **19**, 243–254
JANSSEN, H. A. (1895). *Z. Ver. Dtsch. Ing.*, **39**, 1045–49
JENIKE, A. (1964). University of Utah Bulletin No. 26, Vol **23**, 12
JENIKE, A. W. (1967). *J. of Structural Division, ASCE*. Vol. **93**, No. ST1, 27
JENIKE, A. W. and JOHANSON, J. R. (1969). Trans of the ASME. *J. of Eng. For Ind.*, **91** (2), 339–344
JENIKE, A. W. and JOHANSON, J. R. (1986). Flow of solids. *Newsletter* (Jenike & Johanson) Vol. **VI**, No. **1**. p. 3
JENIKE, A. W., JOHANSON, J. R. and CARSON, J. W. (1973). Trans. ASME. *J. of Eng. for Ind.* **95** (1), pp. 1–16

JUNIOR, C. C. (1983). *2nd Int. Conf. on Design of Silos for Strength and Flow*. 260–271. UK Powder Advisory Centre, London
KVAPIL, R. (1965). *Int. J. Rock Mech. Min. Sci.* **2**, 25–41; 277–304
LIPNITSKII, M. E. and ABRAMOVITSCH, G. R. (1967). *Zhelezobetonnye Bunkera i Silosy, Izd. Lit. po Stroitelstvu*, 2nd edn. Leningrad, p. 208
MCLEAN, A. G. and ARNOLD, P. C. (1976). Trans. ASME, *J. of Eng. for Ind.* **95**, 1370–1374
MURFITT, P. M. (1980). Ph.D thesis, University College London
MURFITT, P. M., BARNSBY, P. L. and NIENOW, A. W. (1981). *Proc. of the 1981 Powtech Conf.*, Inst. of Chem. Engrs. Pub. Series No. 16, pp. D3/S/1 – D3/S/22
NEDDERMAN, R. M. (1981). *Bulk Solids Handling*, **1** (1) 25–30
OOMS, M. and ROBERTS, A. W. (1985). *Bulk Solids Handling*, **5**, 1009–1016
RAVENET, J. (1981). *Bulk Solids Handling*, **1** (4), 667–679
REIMBERT, N. A., Trans. Tech. Publication, Series on Bulk Materials Handling, 1975/76, Vol. 1, No. 3
RICHARDS, P. C. (1977). Trans. ASME. *J. of Eng. for Ind.* **99** (4), 809–813
ROBERTS, A. W. and OOMS, M. (1983). *2nd Int. Conf. on Design of Silos for Strength and Flow*. pp. 151–170. UK Powder Advisory Centre, London
ROTTER, J. M. (1983). *2nd Int. Conf. on Design of Silos for Strength and Flow*. pp. 446–463. UK Powder Advisory Centre, London
SCHWEDES, J. (1985). *German Chem. Eng.*, **8**, 131–138
SOKOLOVSKI, V. V. (1965). *Statics of Granular Media*. Pergamon Press, Oxford
SUGDEN, M. B. (1980). *1st Int. Conf. on Design of Silos for Strength and Flow*. Univ. of Lancaster UK. Powder Advisory Centre, London
SUNDRAM, V. and COWIN, S. C. (1979). *Powder Technol.* **22**, 23–32
VAN ZANTEN, D. C. and MOOIJ, A. (1977). Trans. ASME. *J. of Eng. for Ind.* **99** (4), P2, 814–818
VAN ZANTEN, D. C., RICHARDS, P. C. and MOOIJ, A. (1977). Trans. ASME. *J. of Eng. for Ind.* **99** (4), P3 819–823
WALKER, D. M. (1966). *Chem. Eng. Sci.* **21**, 975–997
WALKER, D. M. and BLACHARD, M. H. (1967). *Chem. Eng. Sci.*, **22**, 1713–1745
WALTERS, J. K. (1973). *Chem. Eng. Sci.* **28**, 13–21
WILLIAMS, J. C., HEAD, J. M. and AHUMADA, J. J. (1983). *2nd Int. Conf. on Design of Silos for Strength and Flow*. pp. 401–423. UK Powder Advisory Centre, London
WOOD, J. G. M. (1983). *2nd Int. Conf. on Design of Silos for Strength and Flow*. pp. 132–145. UK Powder Advisory Centre, London

Symbols

A cross-sectional area of bin
C circumference of bin
C_d overpressure factor for eccentric discharge
D bin diameter
g acceleration due to gravity
H bin height
H_e grain height above the point of effective transition for funnel-flow bins
h grain height
h_0 height of the converging section of the silo
h_6 height of cone of material formed at the top of bin
h_s height at which maximum pressure peak occurs in funnel flow
i = 1 for axisymmetric flow
      = 0 for plane flow
K ratio of lateral to normal pressure in the bin

Pressure profiles in bulk solids storage vessels 47

K_a constant defined by Equation 2.39
K_1 constant given by Equation 2.29
K_2 constant given by Equation 2.33a
K_3 constant given by Equation 2.33b
K_4 constant given by Equation 2.40
K_h constant defined by Equation 2.18
M constant defined by Equation 2.22
N constant defined by Equation 2.21
P_v vertical pressure within the fill
P_w lateral wall pressure
P_{vo} vertical surcharge pressure at the top of hopper
P_h horizontal pressure
Q constant given by Equation 2.20
q dimensionless normal force at the point of effective transition for funnel flow given by Equation 2.45
R constant given by Equation 2.17 and radius of Mohr circle
S_0 constant defined by Equation 2.19
S_w shear stress at the wall
W rate of discharge
X constant defined by Equation 2.46
Y constant defined by Equation 2.47
Z constant defined by Equation 2.24

α hopper half angle
β angle between the major principal stress and the normal to the hopper wall
γ angle of repose for solids
δ effective angle of internal friction
θ angle between the moving channel and the vertical for funnel flow
μ coefficient of wall friction ($\tan\varphi_w$)
ν Poisson ratio for solids
ρ_b bulk solids density
φ angle of internal fiction
φ_w angle of wall friction

Example

For the mass flow silo depicted in Figure 2.26 determine wall pressure profiles during discharge. Bulk material properties are:

δ = 40°
φ_w = 13°
ρ_b = 600 kg/m
K = 0.62

Solution

From the geometry of the hopper:

$$h_0 = \frac{3}{\tan 20} = 8.24 \text{ m}$$

48 Pressure profiles in bulk solids storage vessels

(i) wall pressure profile in the vertical (bin) section:

Assuming fairly smooth and perfectly straight walls, Janssen's equation (Equation 2.6) may be used to give:

$$P_w = KP_v = \frac{D}{4} \frac{g}{\tan\varphi_w} [1 - \exp(-4\tan\varphi_w h/D)]$$

$$= \frac{6}{4} \times \frac{600 \times 9.81}{\tan 13} [-\exp(-4\tan 13 h/6)]$$

which reduces to:

$$P_w = 3.8 \times 10^3 [1 - \exp(-0.15h)] \text{ N/m}^2$$

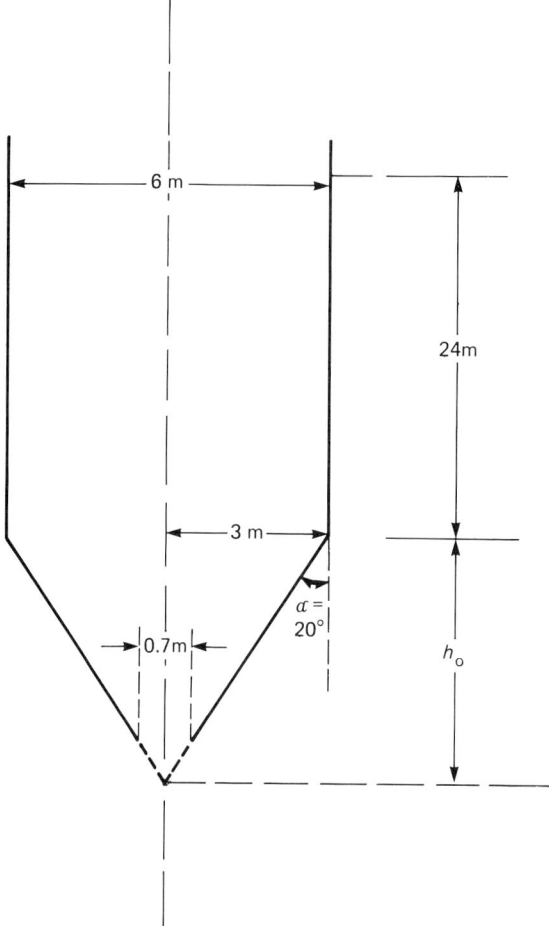

Figure 2.26 Cylindrical silo configuration.

Pressure profiles in bulk solids storage vessels

thus P_w may be evaluated for a series of values of height, h, measured from the top surface. The results are tabulated below:

h(m)	0	2	4	6	8	10	15	20	24
P_w kN/m²	0	10	17	23	27	30	34	36	37

(ii) wall pressures in the converging (hopper) section:

The relevant parameters that must be calculated first are:

$$2\beta = \varphi_w + \sin^{-1}\left(\frac{\sin \varphi_w}{\sin \delta}\right)$$

$$= 20 + \sin^{-1}\left(\frac{\sin 13}{\sin 40}\right) = 40.5$$

$$K_1 = \frac{P_w}{P_v} = \frac{1 + \sin \delta \cos 2\beta}{1 - \sin \cos (2\alpha + 2\beta)}$$

$$= \frac{1 + \sin 40 \cos 40}{1 - \sin 40 \cos 80} = 1.68$$

$$K_3 = K_1 \frac{\tan \alpha + \tan \varphi_w}{\tan 20}$$

$$= 1.68 \frac{\tan 20 + \tan 13}{\tan 20} = 2.75$$

$$K_2 = 2(K_3 - 1) = 3.5$$

Furthermore, at the top of the hopper (bottom of the bin), the vertical pressure, p_{vo}, is calculated using Janssen's equation:

thus:

$$P_{vo} = P_{wo}/K$$

$$= 37/0.62 \text{ at } h = 24 \text{ m}$$

$$= 60 \text{ kN/m}^2$$

Equation 2.34 may be rewritten as:

$$P_v = \frac{\rho_b g h_0}{K_2 - 1}\left[\frac{h}{h_0}\right] + \left[\frac{h}{h_0}\right]^{K_2}\left[P_{vo} - \frac{g h_0}{K_2 - 1}\right]$$

thus:

$$P_v = \frac{600 \times 9.81 \times 8.2}{2.5 \times 10^3} \left(\frac{h}{8.2}\right) + \left(\frac{h}{8.2}\right)^{3.5} \times$$

$$\left[60 - \frac{600 \times 9.81 \times 8.2}{2.5 \times 10^3}\right]$$

which reduces to:

$$P_v = 2.3h + 0.026(h)^{3.5} \quad \text{kN/m}^2$$

and

$$P_w = K_1 p_v = 3.9h + 0.044(h)^{3.5} \quad \text{kN/m}^2$$

thus P_w may be calculated for a series of values of height noting that in this case, height is measured from the vertex of the hopper. The results are tabulated below:

h(m)	1	2	3	4	5	6	7	8
P_w kN/m²	4	8	14	21	32	47	67	94

Wall pressures for the vertical and the converging sections are plotted in Figure 2.27.

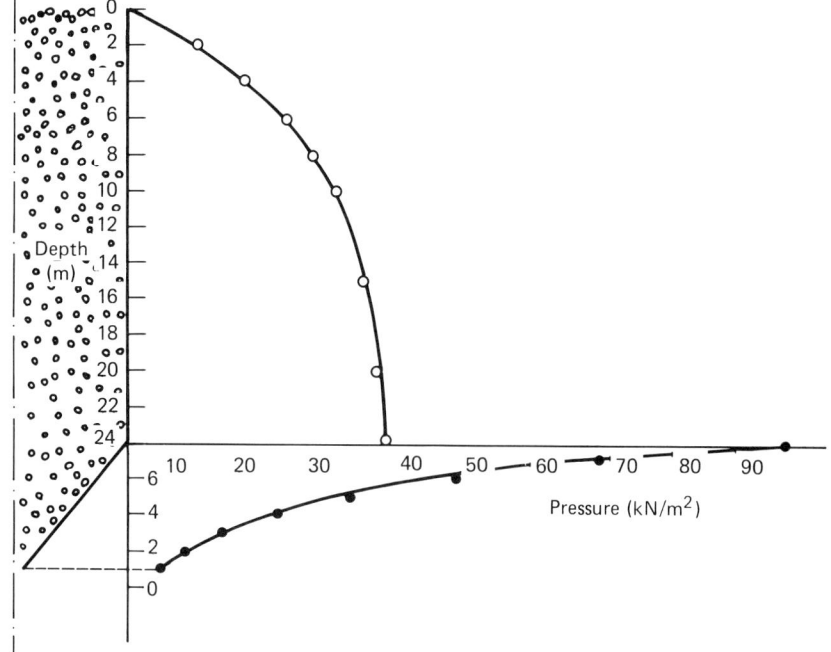

Figure 2.27 Wall pressure profile during discharge.

Chapter 3

Design of storage vessels for particulate solids

Introduction

Storage of granular bulk solids is an important unit operation in the chemical and process industries. Indeed, it is difficult to find any particulate solid that does not pass through a storage device at some stage during its processing. Silos are used commonly not just for stockpiling of raw materials, intermediates and final products, but also as a means of damping out fluctuations in flow and composition within a process.

The rational design of silos for reliable discharge was first developed and published over two decades ago (Jenike, 1961, 1964), and since then the subject has received a sustained interest in many branches of engineering. But despite this it is still thought by many process engineers that a storage silo for bulk solids is nothing more than a square or circular box with a dished bottom having a hole in the middle or on the side of it and a cover at the top. Consequently, the normal design precautions that are commonly observed for other items of equipment in the plant such as reactors and heat exchangers are not taken when it comes to designing storage devices for granular materials. This lack of appreciation often results in severe flow problems during plant operation with considerable loss in production and cost to the industry; the frequency of occurrence and the cost of correcting flow problems in processes that handle solids are substantially higher than those using fluids.

Thus, this chapter concentrates on the design of storage silos for reliable discharge of bulk materials; both mass-flow and funnel-flow silos are considered. The analysis is based largely on the pioneering work of Jenike and his co-workers (Jenike et al., 1959; 1961, 1962, 1964).

In designing for mass-flow silos, the two essential considerations are that the discharge orifice be large enough to prevent the formation of cohesive arches and that the walls of the hopper be steep enough to ensure that particles slide along the walls (Figure 3.1(a)).

In the case of funnel-flow silos, the criterion is to determine the minimum size of the outlet for no arching and no piping (Figure 3.1(b)).

For uninterrupted discharge to occur in either a mass-flow or funnel-flow silo, the design must prevent the formation of stable arches particularly at the outlet; such an obstruction could stop the flow completely. The two basic causes for the formation of a stable arch at the outlet are by the

52   Design of storage vessels for particulate storage

mechanical interlocking of large ($d_p > 3000$ μm) free-flowing particles or by the cohesive bulk strength of fine grains (Figure 3.2).

Mechanical arching may be avoided by having an outlet size that is several times larger than the particle size. Cohesive arching, which is the subject of this chapter is, on the other hand, more common in industrial silos, and its prevention in a given situation requires detailed information on material bulk and frictional properties and the type of flow prevailing in the silo. The concept of a flow/no flow criterion which is explained below is fundamental to Jenike design philosophy proposed to prevent cohesive arching and to ensure reliable flow.

Mass-flow silos: flow/no flow criterion

Consider the case of gravity discharge of a particulate solid in a mass-flow silo with a conical base (Figure 3.3).

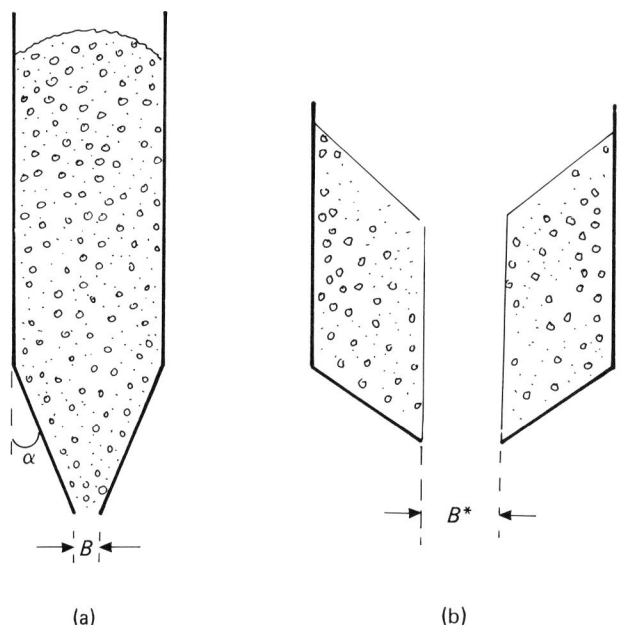

Figure 3.1 Critical dimensions for (a) mass-flow and (b) funnel-flow silos.

Figure 3.2 Arch formation over silo openings; (a) mechanical, (b) cohesive.

Design of storage vessels for particulate storage 53

The pressure p acting in the fill and at the walls increases exponentially with depth up to the transition to the hopper. At this point, there is an abrupt change in pressure due to the switch in the pressure field. Beyond the point of transition, the pressure decreases until it reaches zero at the vertex.

Under this prevailing profile, the degree of compaction, and as a result bulk density and strength of the granular material, will also vary with grain height. Table 3.1 shows the experimentally observed variation with depth of fill for a few materials (Ravenet, 1983). The variation in bulk material strength f with grain height is similar to that of the pressure in the fill (Figure 3.3).

With the increase in the cohesive strength of the powder as a result of consolidation, the tendency for the formation of a stable structure, e.g. an

Table 3.1 Variation in bulk solids density with depth in a 12 m diameter silo (Ravenet, 1983)

Material	Depth of fill (m)	Bulk density (kg/m^3)
Wheat	15	830
	30	1050
Cement	15	1430
	30	2000
Sand	15	1800
	30	2400

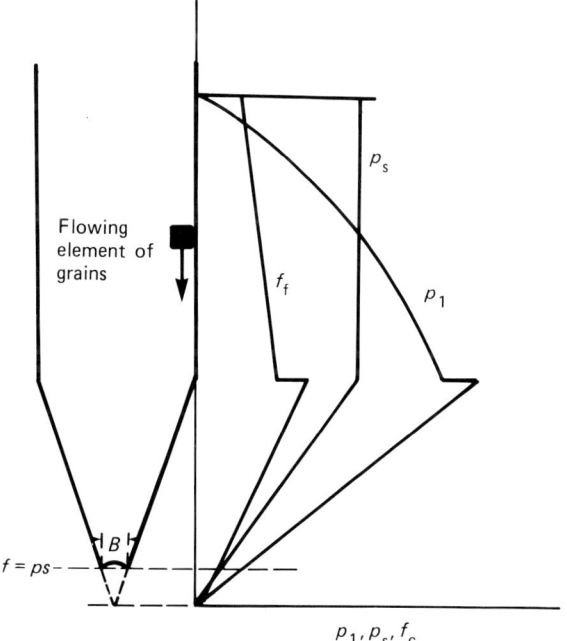

Figure 3.3 Critical silo opening for no arching.

54 Design of storage vessels for particulate storage

arch or a dome, also increases. However, in determining the critical hopper dimensions for uninterrupted flow, it is the relation between the bulk strength that a powder develops at an exposed surface, (i.e. its unconfined yield strength, f_c) and the consolidating pressure ($\bar{p}_1$) that is of primary significance.

The relation between f_c and $\bar{p}_1$ may be explained by considering a cylindrical specimen of the powder uniformly compacted by a vertical (major) consolidating pressure, $\bar{p}_1$, in a container with practically frictionless walls (Figure 3.4). After compaction, the container is removed carefully to ensure that the sample is not disturbed and the vertical compressive load required to just crush the sample is obtained; this is equal to the unconfined yield strength of the powder, f_c. The experiment may be repeated for several values of consolidating pressure and in each case the corresponding value of the unconfined yield strength, f_c, is obtained. A plot of f_c against $\bar{p}_1$ is known as the flow function (FF) for the powder (Figure 3.5).

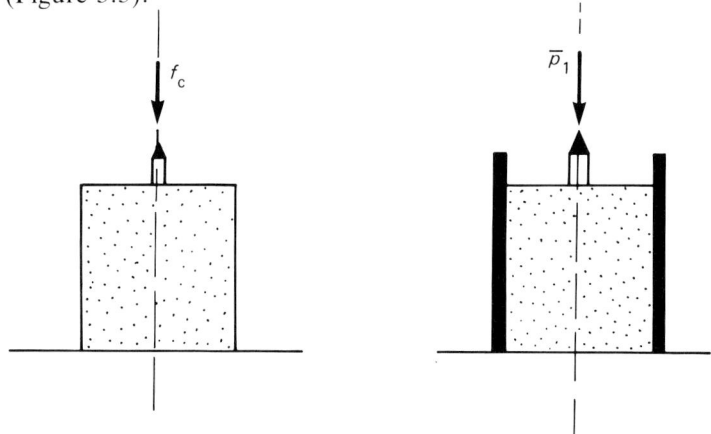

Figure 3.4 Unconfined yield strength of cohesive materials.

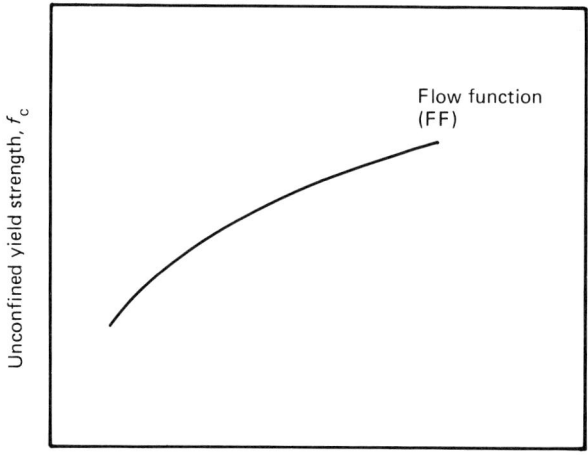

Figure 3.5 Material flow function.

Design of storage vessels for particulate storage   55

Evidently the flow function, FF, is a property of the bulk material and its degree of compaction. Jenike used this ratio to classify powders. Table 3.2 shows the results of his investigation.

Schubert (1979; 1984) extended Jenike's classification to include powders with a flow function less than 1.0; such a 'hardened' material is capable of forming arches in a silo with a very large span.

In practice, the variation of f_c with $\bar{p}_1$ is obtained by a specially designed shear cell device, developed by Jenike (1964). This is described later in this chapter.

Thus, a stable cohesive arch is formed when the unconfined yield strength of the powder exceeds the forces tending to break it. Jenike (1961; 1962) and Walker (1966) assumed that the only force acting on a thin uniform layer over an arch, which tends to break it, is its own weight. This implies that the mass of the material above the arch exerts no force upon it, and thus gives the worst condition to maintain flow.

Figure 3.6 shows the equilibrium of forces acting in such an arch of material of uniform thickness, t. At the point of incipient motion, the Mohr semicircle representing the stresses on the surface of the arch passes through the origin (zero shear and normal stress on the surface) and touches the material yield locus (Figure 3.7). On the Mohr semicircle, point A represents the maximum shear stress that the arch can support; from the geometry of the Mohr circle, this maximum shear force, $S(\max)$, is half the unconfined yield strength of the powder. Thus;

$$S(\max) = f_c/2 \tag{3.1}$$

Table 3.2 Jenike's classification of powders by means of their flow function, FF

FF < 2	Very cohesive (non-flowing)
2 < FF < 4	Cohesive
4 < FF < 10	Slightly (easy) flowing
10 < FF	Free flowing

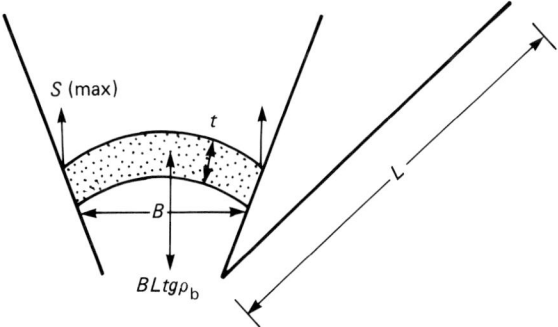

Figure 3.6 Vertical equilibrium of forces acting on an arch.

56 Design of storage vessels for particulate storage

When the arch is on the verge of collapsing, its weight is just balanced by the maximum shear stress, $S(max)$, developed at the boundary between the grains and the walls. In the case of a wedge-shaped hopper with length L and span B (Figure 3.6), the vertical force balance on the element gives:

$$BLt\rho_b g = S(max)2Lt \tag{3.2}$$

that is

$$B = 2S(max)/\rho_b g \tag{3.3}$$

Substituting for $S(max)$ from Equation 3.1, and noting that for the arch to collapse the major principal stress, p_s, acting along the surface of the arch must be at least equal to the unconfined yield strength, gives:

$$B = f_c/\rho_b g = p_s/\rho_b g \tag{3.4}$$

Equation 3.4 indicates that the magnitude of the major stress acting along the arch surface is directly proportional to its span B (Figure 3.3).

For a conical hopper, a similar force balance at the point of incipient motion yields:

$$B = f_c/2\rho_b g = p_s/2\rho_b g \tag{3.5}$$

Thus in general;

$$p_s/\rho_b g B = 1/(1 + i) \tag{3.6}$$

where $i = 0$ for wedge-shaped silos
$i = 1$ for conical hoppers

Equation 3.6 was subsequently modified to Equation 3.7 by Jenike and Leser (1963) to allow for the variation in the thickness of the arch:

$$p_s/\rho_b g B = 1/H(\alpha) \tag{3.7}$$

where $H(\alpha)$ is only a function of the hopper half-angle, α, and silo geometry (Figure 3.8); these lines may be expressed adequately by the following equation (Arnold and McLean 1976a,b, Arnold et al., 1980).

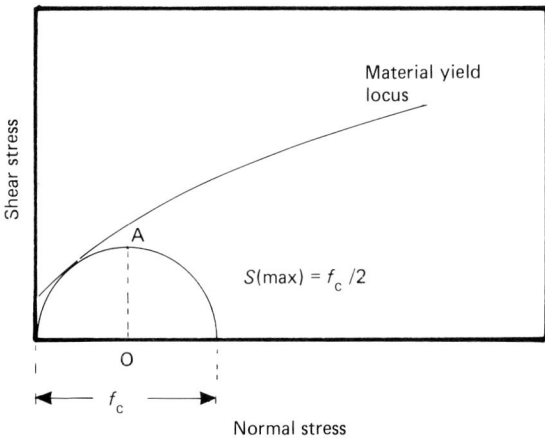

Figure 3.7 Stresses at a free arch.

$$\frac{1}{H(\alpha)} = \left[\frac{65}{130 + \alpha}\right]^i \left[\frac{200}{200 + \alpha}\right]^{1-i} \quad (3.8)$$

with α in degrees and

$i = 1$ for circular and square openings
$i = 0$ for slot opening ($L \geq 3B$)

In Figure 3.3, the point at which the major principal stress curve intersects the bulk material strength curve satisfies Equation 3.4 and is used to determine the critical hopper opening needed to prevent the formation of stable cohesive arches in the silo; below this point, the powder has sufficient cohesion to support an arch and, consequently, no flow occurs. Above the point, the bulk material has insufficient strength to support an arch, and thus the grains flow. This is the concept of the flow/no flow criterion.

Noting from Figure 3.9 that $B = 2r \sin \alpha$, Equation 3.4 may be rewritten as:

$$p_s = 2\rho_b g r \sin \alpha \quad (3.9)$$

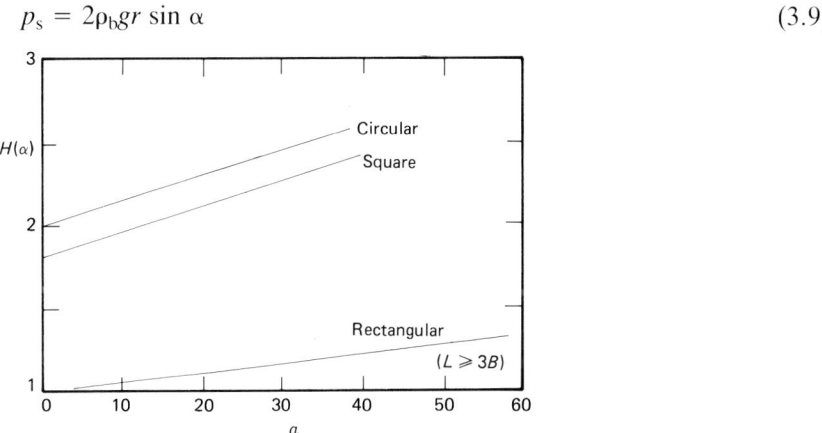

Figure 3.8 Function of $H(\alpha)$ (Jenike, 1964).

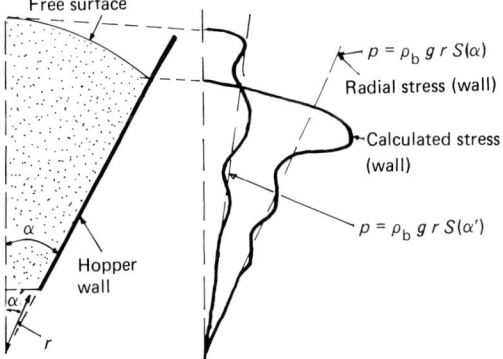

Figure 3.9 Radial stress and calculated stress (Johanson and Colijn, 1964) for a converging channel.

58 Design of storage vessels for particulate storage

which is the magnitude of the major principal stress trying to break the arch.

Furthermore, in the vicinity of the hopper outlet, Jenike (1962) showed that the pressure field is particularly radial and that the mean pressure at the wall is given by:

$$p = \rho_b g r S(\alpha) \qquad (3.10)$$

where $S(\alpha)$ is the radial stress function at the wall for the converging channel (Figure 3.9). During discharge, the Mohr semicircle representing the stresses in the powder is always tangential to the effective yield locus and decreases in diameter as the powder moves downwards (Figures 3.10; 3.11). From the geometry of the Mohr semicircle (Figure 3.10);

$$\bar{p}_1 = OC = OA + AC = OA + AP = OA(1 + \sin \delta) \qquad (3.11)$$

and

$$\bar{p}_2 = OB = OA - AB = OA - AP = OA(1 - \sin \delta) \qquad (3.12)$$

Hence:

$$\bar{p}_1/\bar{p}_2 = (1 + \sin \delta)/(1 - \sin \delta) \qquad (3.13)$$

Noting that

$$OA = p = (\bar{p}_1 + \bar{p}_2)/2 \qquad (3.14)$$

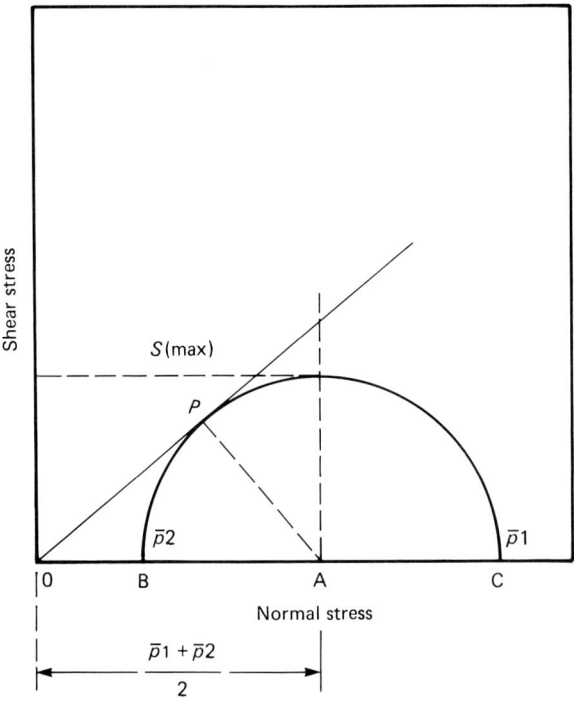

Figure 3.10 Typical Mohr semicircle diagram.

Design of storage vessels for particulate storage   59

thus:

$$\bar{p}_1 = p(1 + \sin \delta) = \rho_b g r S(\alpha)(1 + \sin \delta) \tag{3.15}$$

Equation 3.15 gives the magnitude of the major principal stress acting on the arch during its formation. For each $\bar{p}_1$, the unconfined yield strength is obtained from the flow function plot.

Dividing Equation 3.15 by Equation 3.9 gives:

$$\bar{p}_1/p_s \leq S(\alpha)(1 + \sin \delta)/2 \sin \alpha = \text{ff} \tag{3.16}$$

For a given powder and silo, δ and α are constant, so that the ratio $\bar{p}_1/p_s$, known as the flow factor (ff), is a constant: it is the ratio of the major principal stress acting on the arch during its formation (hence determining its strength) to the major principal stress trying to break the arch. At the point of failure, of course, $p_s = f_c$ and thus ff = FF = $\bar{p}_1/f_c$.

Enstad (1975) developed an analytical approximation for $S(\alpha)$ for wedge-shaped silos. Arnold and McLean (1976a,b) extended this theory to include axisymmetric as well as plane-flow silos. For detailed derivation of the equations, the original papers may be consulted; for design purposes, their final expression for the mean stress p (Equation 3.10) is given by:

$$p = \frac{\rho_b g Y r}{X - 1} + \left[p_R - \frac{\rho_b g R}{X - 1} \right] \{r/R\}^X \tag{3.17}$$

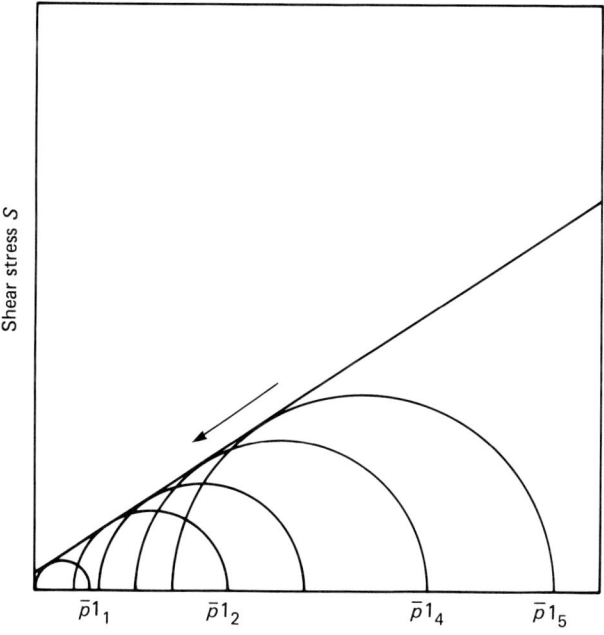

Figure 3.11 Tangential relationship between Mohr semicircles and effective yield locus.

where

$$X = \frac{2^i \sin \delta}{1 - \sin \delta} \left\{ \left[\frac{\sin[2\beta + \alpha]}{\sin \alpha} \right] + 1 \right\} \quad (3.18)$$

and

$$Y = \frac{[2(1 - \cos(\beta + \alpha))]^i \sin \alpha (\beta + \alpha)^{1-i} + \sin \beta \sin^{1+i}(\beta + \alpha)}{(1 - \sin \delta) \sin^{2+i}(\beta + \alpha)} \quad (3.19)$$

As $r \to 0$, i.e. in the vicinity of the outlet, Equation 3.17 reduces to:

$$p = \frac{\rho_b g Y r}{X - 1} \quad (3.20)$$

which indicates that near the outlet the stress field has a radial distribution.

Comparing Equation 3.20 with Jenike's expression (Equation 3.9) yields:

$$S(\alpha) = \frac{Y}{X - 1} \quad (3.21)$$

Substituting in Equation 3.16 gives:

$$ff = \frac{Y(1 + \sin \delta)}{2(X - 1)\sin \alpha H(\alpha)} \quad (3.22)$$

The function $H(\alpha)$ accounts for variation in the thickness of the arch (Jenike and Leser, 1963). Arnold and McLean provided Equation 3.8 for $H(\alpha)$ based on a chart solution given by Jenike and Leser (Figure 3.8).

Arnold and McLean (1976a,b) found excellent agreement between calculated ff values (Equation 3.22) and chart values provided by Jenike (1961) and Johanson and Jenike (1962).

Jenike's chart solutions for ff based on computed values of $S(\alpha)$ for different effective angles of friction, δ, angles of wall friction, φ_w, and hopper half-angle, α, are shown in Figures 3.12–3.15 for cylindrical and slot silos. Figures 3.12 and 3.13 give the transition between mass flow and funnel flow. For conical hoppers, the bounds may be described by the following theoretical equation (Jenike, 1961):

$$\alpha = \frac{\pi}{2} - \frac{1}{2} \cos^{-1}\left[\frac{1 - \sin \delta}{2\sin \delta} \right] - \beta \quad (3.23)$$

where β is given by Equation 3.24:

$$2\beta = \varphi_w + \sin^{-1}\left[\frac{\sin \varphi_w}{\sin \delta} \right] \quad (3.24)$$

For slot silos, the bounds in Figure 3.13 are described adequately by the following empirical equation (Arnold and McLean, 1980):

Design of storage vessels for particulate storage 61

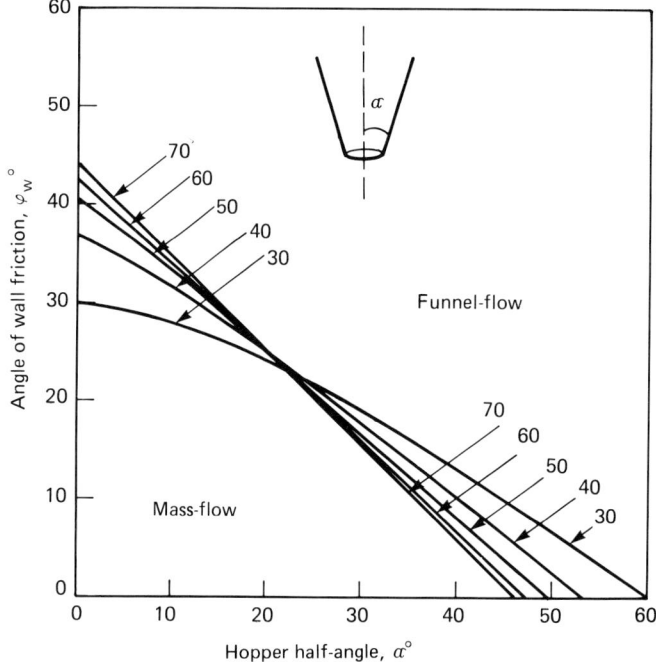

Figure 3.12 Funnel-flow/mass-flow bounds for axisymmetric (conical) silos (Jenike, 1964).

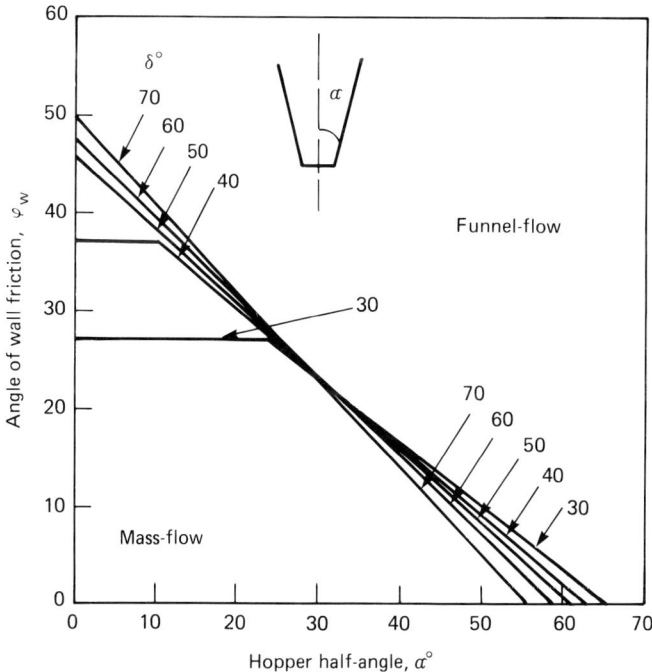

Figure 3.13 Funnel-flow/mass-flow bounds for plane (slot) silos (Jenike, 1964).

62 Design of storage vessels for particulate storage

$$\alpha = \frac{\exp[3.75(1.01)^{(\delta - 30)/10}] - \varphi_w}{0.725(\tan \delta)^{1/5}} \tag{3.25}$$

for $\varphi_w < \delta - 3$

In Equation 3.25 φ_w and δ are in degrees

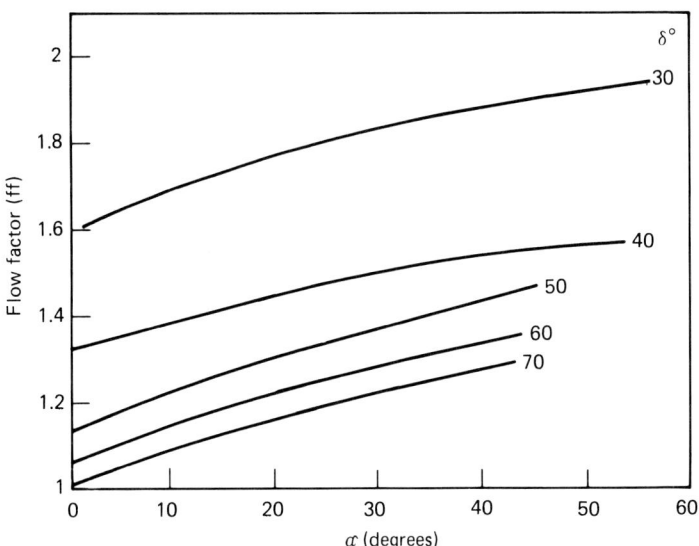

Figure 3.14 Flow factor (ff) for axisymmetric (conical) mass-flow silos.

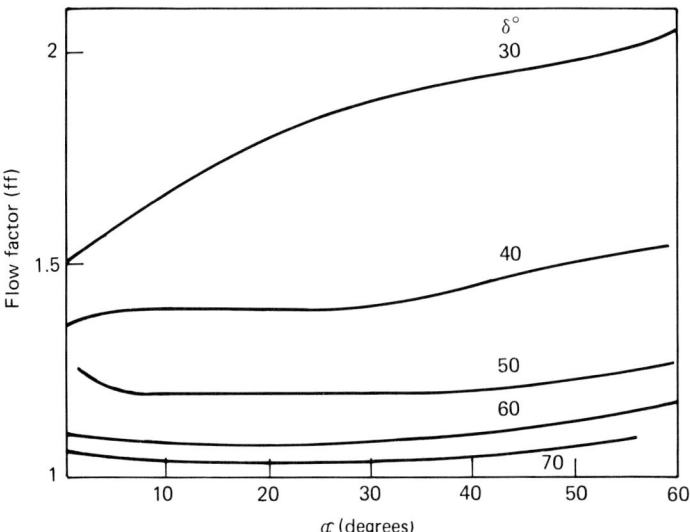

Figure 3.15 Flow factor (ff) for plane (slot) mass-flow silos.

Design of storage vessels for particulate storage 63

It is important to note that the curves in Figures 3.12 and 3.13 indicate that a small change in φ_w is enough to change a mass-flow to a funnel-flow silo.

Figure 3.16 shows the flow factor (ff) superimposed on the flow function (FF) for the material under investigation. The critical condition for uninterrupted discharge occurs at the intersection between the two lines; this defines the critical unconfined yield strength, $f_{c\,(crit)}$ which is used to determine the minimum hopper opening given by Equation 3.26:

$$B = \frac{H(\alpha)f_{c(crit)}}{\rho_b g} \tag{3.26}$$

Of course, there is no reason for the two lines to intersect and, in practice, three basic possibilities exist:

(i) The flow function (FF) lies entirely below the flow factor (ff) (Figure 3.17), that is, the applied stress exceeds the unconfined yield strength. Consequently, the minimum size of the opening is not determined on the basis of cohesive arching; design must, however, prevent the formation of mechanical arching and should provide the desired rate of discharge.

(ii) The FF curve lies entirely above the ff line (Figure 3.18), that is, material unconfined strength exceeds the applied stress. Consequently, gravity unloading is impossible and the design must be repeated by selecting a new hopper half-angle and, if possible, by using a smoother wall material.

(iii) The FF curve intersects the ff line (Figure 3.16). The minimum hopper opening to maintain flow is evaluated using the point of intersection and Equation 3.26.

The design of the silo should also allow for the fact that the flow functions (FF) for most commercial materials exhibit pronounced time-dependency due to consolidation (Figure 3.19), and vary with particle size and moisture content (Figures 3.20, 3.21).

With regards to the flow factor (ff), Borg (1982) investigated the flow properties of 40 synthetic rubbers by adding small amounts (2–5%) of flow aids. He observed improvements in the instanteaneous flowability by a

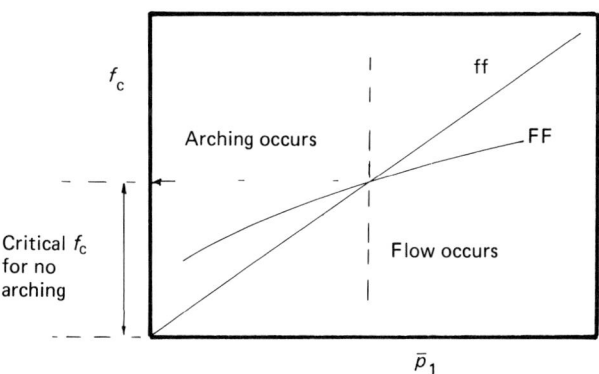

Figure 3.16 Critical f_c for no arching.

64 Design of storage vessels for particulate storage

factor of 2.0 to 3.0, and by a factor of 1.5 to 2.0 for storage after one day; an improvement in the flowability factor of 2.0 results in a reduction of 60–70% in the minimum hopper opening.

It is therefore imperative that the flow function (FF) and the flow factor (ff) representing the material under the actual storage conditions are used in design. It is equally essential to maintain these conditions for as long as the plant is in operation. This is particularly important in the storage of biological products such as foodstuffs. With these materials, environmental conditions such as relative humidity, temperature, and oxygen concentration could support the growth of micro-organisms and biochemical reactions at rates high enough to alter the chemical and physical composition (and thus ff and FF) of the product during storage. Therefore, although the basic design philosophy is essentially the same for both biological and non-biological bulk solids, the fact that biological products are not inert requires the process engineer to specify the optimum environmental conditions to prevent product degradation during storage.

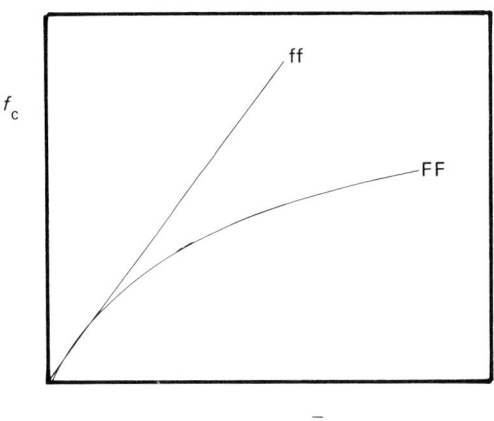

Figure 3.17 ff/FF condition for cohesive arching to form.

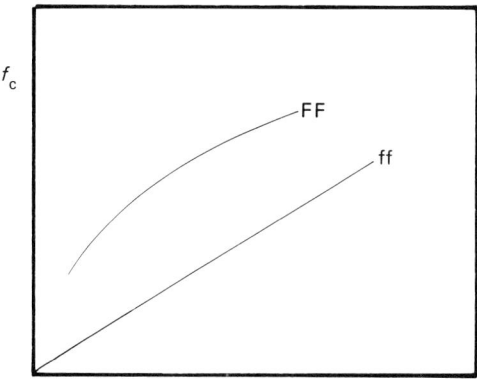

Figure 3.18 ff/FF condition for no gravity flow.

Design of storage vessels for particulate storage 65

In summary, the design of a mass-flow silo for any particulate solid requires information on the following properties of the bulk material:

1. Effective angle of friction of the powder, δ;
2. The flow function of the powder, FF;
3. The flow factor of the powder and the silo, ff;
4. Angle of wall friction, φ_w.

If the above parameters show time-dependency during storage, then time-consolidated results should be obtained and used in design.

In most cases, the above information is generated experimentally by use of a direct shear cell device developed and tested originally by Jenike (1964). This is described below.

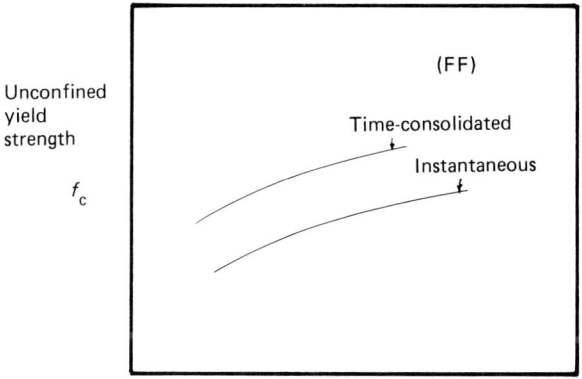

Figure 3.19 Instantaneous and time-consolidated flow functions.

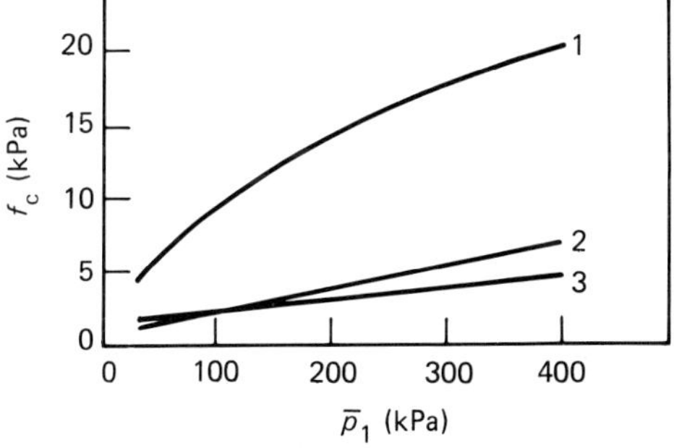

Figure 3.20 FF curves for coal (Tamura and Haze, 1985).

Key	Particle size max. (mm)	Free moisture content (%)
1	3	14.3
2	50	14.2
3	10	13.6

Measurement of material properties for silo design

The basic shear cell, called the Flow factor tester, is shown schematically in Figure 3.22. It consists of a split circular cell with a fixed base, a shearing ring and a lid. The cell that is most commonly used has an inner diameter, d_c, of 0.0953 m (cross-sectional area, $A = 0.007$ m^2); a smaller cell, $d_c = 0.0635$ m, $A = 0.003$ m^2, is also available for studies under high compaction.

The sample is equilibrated to the desired conditions in terms of, for example, moisture content and temperature. The cell is filled with a specimen of the powder, the excess material is scraped off and the lid is positioned on it. The sample is compacted by applying a vertical consolidating load, LC, on the lid under controlled conditions. After compaction, the drive mechanism is switched on; this moves the shearing ring horizontally at a constant rate of 2.3 mm/min, thus shearing the sample at a constant rate of strain.

A load cell and a chart recorder are used to measure and display the shear force needed to maintain the motion of the ring; the recorder output is a shear-stress/time plot, but as the rate of strain is constant, it is also a graph of shear-stress shear strain. For any particulate solids, the shape of

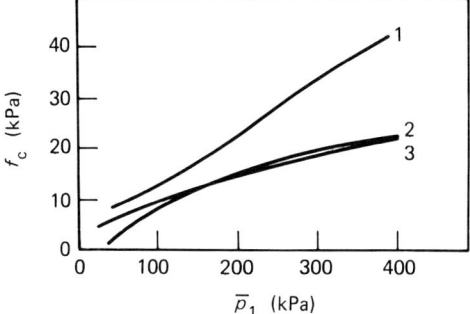

Figure 3.21 FF curves for limestone (Tamura and Haze, 1985).

Key	Particle size max. (mm)	Free moisture content (%)
1	3	4.2
2	20	4.2
3	10	3.5

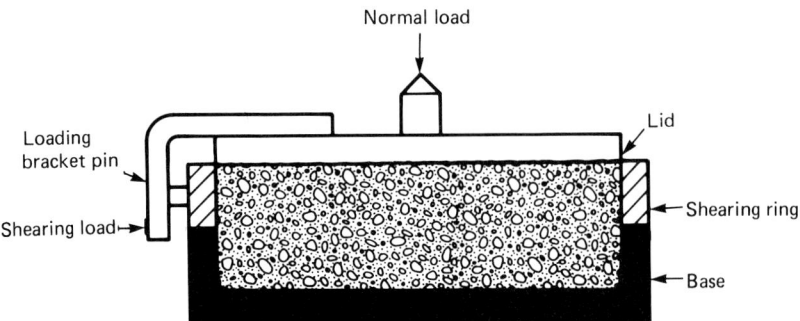

Figure 3.22 Jenike direct shear cell tester.

Design of storage vessels for particulate storage

the curve depends on the degree of compaction, but in general there are three likely possibilities (Figure 3.23) (Williams, 1983; Williams and Birks, 1965).

(i) Type A represents the critical conditions at which the specimen shears with little change in volume; the work done on the sample by the applied shear force is used entirely to shear the sample.

(ii) A tightly compacted sample (type B) dilates during the shearing period; work is done against the applied load.

(iii) A loosely packed sample (type C) contracts during shearing; work is done by the applied load.

The test procedure is in two parts. The first involves sample preparation, while during the second phase the sample is sheared. These are discussed below.

(i) Sample preparation

With the shearing ring offset by about 2.5 mm, a mould ring is put on top of it (Figure 3.24). The equilibrated powder is poured into the cell, the excess material is scraped off and the lid is put on the top. A consolidation load, LC, is applied to the lid, which is then twisted to compact the specimen. Compaction of the powder is carried out under standardized conditions to ensure reproducibility. After consolidation, the load, lid and the mould ring are removed and the excess material is carefully taken off to the level of the shearing ring. The lid is put back on.

The sample is now ready for the second stage.

(ii) Sample shearing

A normal load, $L1$, smaller than the consolidation load, LC, used during sample preparation, is placed on the lid and the sample is sheared until

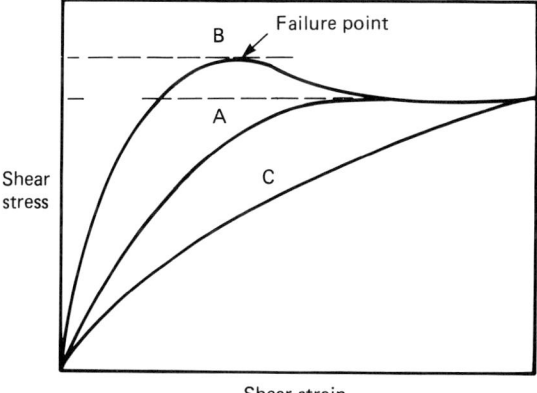

Figure 3.23 Typical stress–strain plots.
Key
A   Critical state
B   Overconsolidate
C   Under compacted

68 Design of storage vessels for particulate storage

failure occurs. The aim at this stage is to obtain a specimen condition that results in a type A behaviour, i.e. failure with no change in volume. If the response curve is of type B or C, the whole experiment is repeated with a different value of the consolidation load, LC, until type A curve is obtained during shearing with the normal load $L1$. The shear stress (shear force/cell area), $S1$, that causes failure under the applied major compressive stress, $p1$, ($L1$/cell area) is obtained from the trace and is plotted on a shear-stress/normal stress graph; the plot is referred to as the material yield (failure) locus (Figure 3.25). The point ($S1, p1$) is particularly important in design; it is the last point on the yield locus and is used to obtain the magnitude of the major principal stress acting on an arch and the effective angle of internal friction of the material; this is discussed below.

The above experimental procedure is repeated with several fresh samples, all prepared in an identical manner described above using the same consolidation load, LC, that yielded the type A curve with the compressive load $L1$. Each sample is then sheared in turn with a different compressive load, $L2, L3, L4$, etc. on the lid; the applied load used in each case must be lower than $L1$, so that the trace for these subsequent samples are all of Type B. For each specimen, the corresponding shear stress, $S2$, $S3$, $S4$, etc. that causes failure is obtained from the chart recorder. The results plotted in Figure 3.25 give a single failure locus for the powder.

In this way, a family of yield loci may be obtained by using different consolidation loads, LC, applied during the initial stage of sample preparation (Figure 3.26).

It is important that the normal loads used during sample preparation and shearing represent typical pressures that exist in the silo during the actual

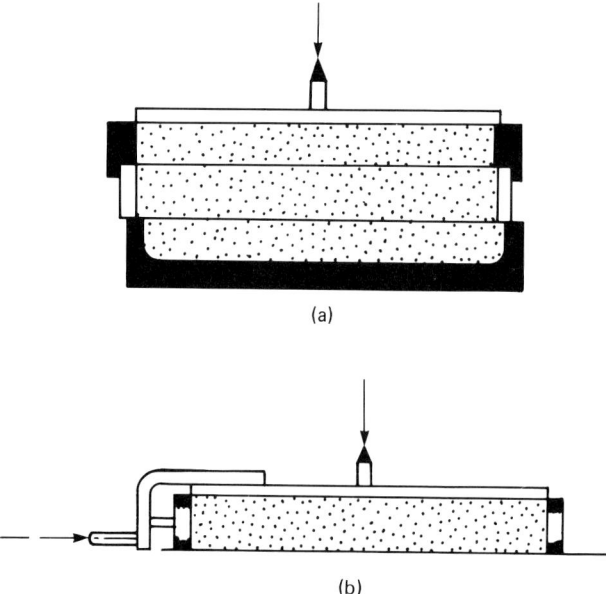

Figure 3.24 (a) Shear cell with mould ring during sample preparation; (b) Shear cell with base removed for measurement of angle of wall friction.

Design of storage vessels for particulate storage   69

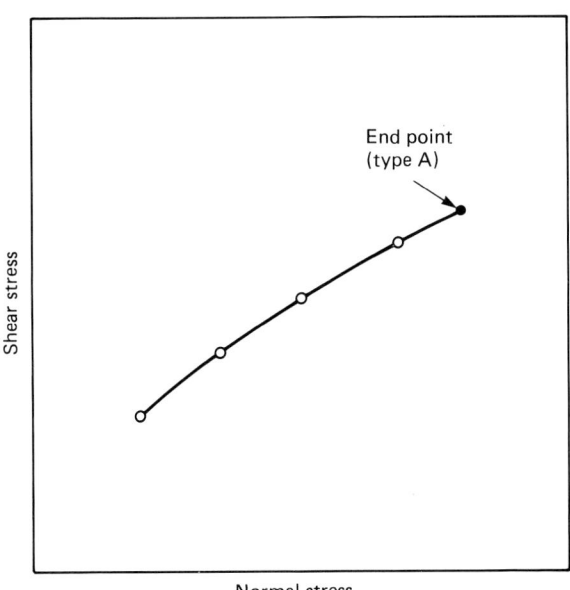

Figure 3.25 Typical material yield locus.

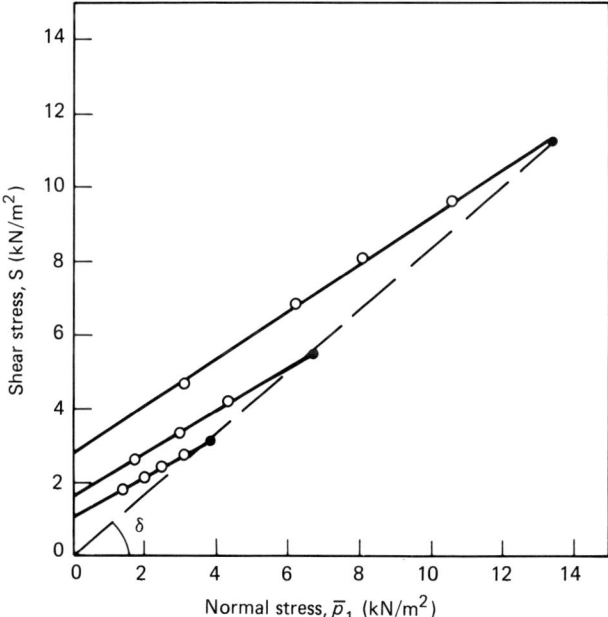

Figure 3.26 Family of yield loci and effective angle of internal friction.

70 Design of storage vessels for particulate storage

operation. Moreover, for time-consolidated results, the test procedure is modified to allow for this; samples are tested after a period of storage under consolidation (the consolidation time must be equal to the expected hold-up time in the silo).

The yield loci for most commercial products are not straight lines and do not pass through the origin. However, Jenike observed that for most powders, the end-points on the yield loci lie closely on a straight line that passes through the origin. The angle between this line and the abscissa defines the effective angle of internal friction, δ (Figure 3.26).

For each individual yield locus, the Mohr semicircle passing through the last point (the type A curve) and tangent to the locus is used to obtain the major principal stress, $\bar{p}_1$ (Figure 3.27). The unconfined yield strength, f_c, is obtained by drawing the Mohr semicircle passing through the origin and tangent to the yield locus. Intersection of this Mohr semicircle with the normal stress axis defines the unconfined yield strength, fc. Each yield locus provides a single point on the flow function plot (Figure 3.5).

Wall yield locus

The condition at the boundary between the bulk material and the walls of the silo is of particular importance in design for mass flow. The necessary

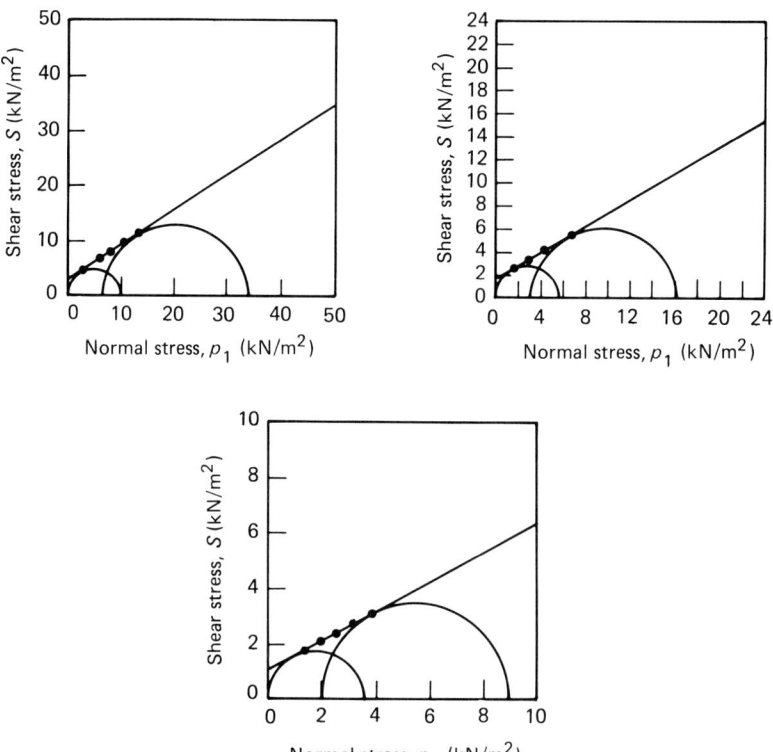

Figure 3.27 Family of yield loci with the tangential Mohr semicircles for f_c and $\bar{p}_1-$.

design information is obtained from the shear cell by replacing the base of the cell with a flat piece of material which is to be used as the construction material for the silo; the shearing ring is thus in direct contact with the wall material (Figure 3.24).

Sample preparation and shearing is as described for material yield loci. A typical plot of the shear stress (S_w) and normal stress (p_w) is shown in Figure 3.28; this is often a straight line that passes through the origin with a slope that defines the coefficient of wall friction, $\mu = \tan\varphi_w$.

Summary of design method for mass-flow silos

The analysis presented so far is summarized below to give a step-by-step procedure to determine the critical hopper half-angle, α, and the minimum size of the outlet for mass flow:

1. Using a Jenike shear cell, develop a family of yield loci (3–4) and present the information on a shear-stress/normal stress plot (Figure 3.26). If the powder consolidates during storage, this plot should be for time-consolidated material.

2. For each individual yield locus:
(i) draw a Mohr semicircle passing through the origin and tangent to the yield locus; it is often necessary to extrapolate the yield locus for this purpose. Read off the value of the unconfined yield strength, f_c, from the point of intersection of the Mohr semicircle and the abscissa (Figure 3.27).

(ii) draw a Mohr semicircle passing through the last point and tangent to the yield locus, and thus obtain the value of the major principal stress, $\bar{p}_1$; this is the point of intersection of the semicircle and the abscissa (Figure 3.27).

3. Obtain the failure function (FF) for the material by plotting values of f_c against $\bar{p}_1$ (Figure 3.5).

4. Join the end-points of the individual yield locus by the best straight line that passes through the origin and measure its slope; the slope defines the effective angle of internal friction, δ (Figure 3.26).

5. Using the Jenike shear cell, measure the angle of wall friction, φ_w (Figure 3.28).

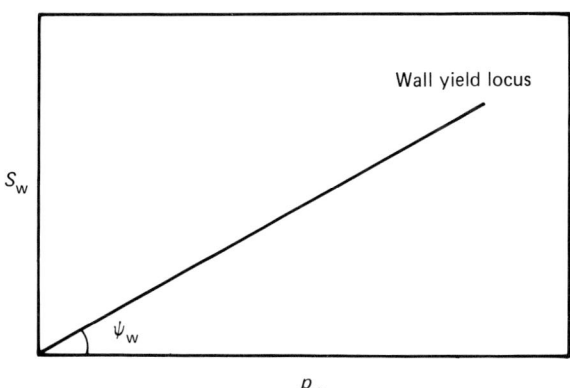

Figure 3.28 Wall shear stress–normal stress plot.

6. For the measured δ and φ_w (steps 4 and 5), calculate the hopper half-angle, α, required to ensure mass flow using Equations 3.23 and 3.25 for axisymmetric and plane-flow geometry respectively. Alternatively, use the appropriate chart (Figures 3.12 and 3.13) for this evaluation.

7. Calculate the flow factor (ff) for the hopper and the material under investigation using Equation 3.25. Alternatively, the appropriate chart (Figure 3.14 and 3.15) may be used for this purpose.

8. On the failure function (FF) plot, draw a straight line passing through the origin with a slope of l/ff. The intersection of the two lines gives the critical unconfined yield strength, $f_{c_{(crit)}}$.

9. Determine the value of the function $H(\alpha)$ using the Equation 3.8.

10. Evaluate the minimum hopper opening for mass flow using Equation 3.26.

Considering that the design method presented in this section was reported well over two decades ago, it is surprising to observe the lack of published information in the open literature on the performance of full-scale silos designed on the basis of Jenike's technique. The few papers that have been published appear to give contradictory conclusions. For example, Walker (1967) concluded that Jenike's approach overestimates the size of the outlet considerably. Wright (1970, 1972), on the other hand, claimed excellent agreement between calculated and observed critical hopper outlets, while Eckhoff and Leversen (1973) reported that Jenike's method overestimates the hopper opening by about 100%.

As far as the prediction of the critical hopper half-angle, α, is concerned, both Wright (1972) and Eckhoff and Leversen (1973) observed an over-design of approximately 8–10 degrees using Jenike's method. However, both publications recommended the technique as a practically wise design.

Eckhoff and Leversen point out that the critical hopper outlet size is extremely sensitive to even small variations in shear cell data, particularly in the region of low normal stresses. They thus recommended that the usual technique of extrapolating the material yield locus into that region should be avoided and that better measuring techniques are needed to obtain the relevant information in the low normal stress region.

It should also be pointed out that a basic assumption in the Jenike design philosophy is that forces due to momentum transfer are small in comparison to body forces. Consequently the method only applies to cases where particle velocity is low, and well below the free-fall velocity. In practice, however, this is not a major drawback, as the outlet of the silo is normally connected directly to a feeder or some other metering or flow-controlling device which restricts the flow rate to well below the free-fall value (Gaylord and Gaylord, 1984; Safarian and Harris, 1984).

Finally, the design of normal mass-flow silos usually results in relatively tall structures that often require expensive filling equipment to lift the particulate. For cohesive materials and in cases where the available space imposes a design restriction, wedge-shaped silos are recommended; for a given duty, the hopper half-angle, α, for plane-flow geometry is larger than that for an axisymmetric hopper. Thus, the plane-flow silo will be shorter than the axisymmetric one (see example at the end of this chapter). Richards (1966) points out that almost any wedge-shaped silo with an $L:B$ ratio of 4 to 5:1 for the opening will give reliable discharge, provided that

Design of storage vessels for particulate storage   73

the width of the silo is large enough. Jenike gives the minimum ratio for length to width for slot silos as 3:1; this is to prevent the occurrence of funnel-flow in the upper region of the silo. In practice, the geometry of the basic silo is often modified to overcome some of these difficulties; the design of transition and Binsert™ silos (Jenike and Johnson, 1986) are good examples of such modifications. Detailed design information on these modified mass-flow silos are not available in the open literature, but neither are reported failures.

Funnel-flow silos

For funnel-flow silos, the outlet size is computed for no arching and no piping.

For no piping, the critical silo outlet dimension is B^*, which is the diameter for a circular silo and the diagonal for a rectangular or square geometry silo (Figure 3.29).

Jenike (1964) proposed the following equation for the evaluation of B^* for no piping (Figure 3.29):

$$B^* = \frac{G(\varphi_t)p_v}{\rho_b g} \tag{3.27}$$

where p_v, the major consolidating stress is obtained from Janssen's formula presented in the last chapter;

$$p_v = \frac{A\rho_b g}{CK\mu}\{1 - \exp[-K\mu(hC/A)]\} \tag{3.28}$$

where K is given by Equation 3.29:

$$K = \frac{1 - \sin\delta}{1 + \sin\delta} \tag{3.29}$$

φ_t is the static angle of internal friction of the material. It is obtained from the slope of the time-consolidated yield locus at the point that is tangential to the Mohr circle passing through the origin. Jenike provided a chart (Figure 3.30) for the function $G(\varphi_t)$.

Arching does not normally occur for axisymetric silos designed to prevent piping. For slot silos, Jenike recommended that the technique

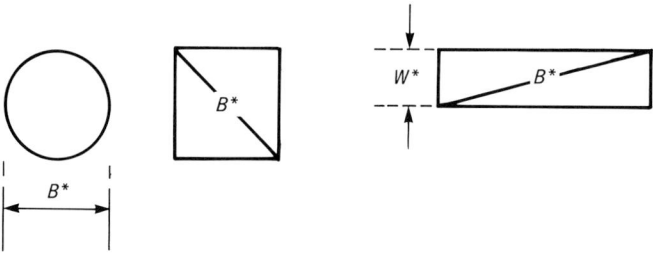

Figure 3.29 Outlet dimensions for funnel-flow silos.

proposed for mass flow may be adopted, but for funnel flow a constant flow factor, ff = 1.7 is used; this corresponds to an angle of 30 degrees for the live channel, θ, (see Figure 2.4 in the previous chapter). The critical width, W^*, for no arching is then given by Equation 3.26, with:

$$H(\alpha) = H(30) = 1.15$$

thus;

$$W^* = \frac{1.15 \, f_{c(crit)}}{g\rho_b} \qquad (3.30)$$

$f_{c(crit)}$ is obtained from shear cell tests as described for mass-flow silos.

Flow-promoting devices

There are many reasons for using feeders and flow-promoting devices to control discharge from silos, to obtain uniform flow from cone (funnel-flow) bins, or to restart a blocked hopper, which is sometimes necessary even with a mass-flow silo. For example, with easily aeratable materials, air is entrained as particles fall freely into the container during loading. Consequently, the material is flowable and may be discharged readily immediately after the filling process. However, if discharging is stopped for any appreciable length of time, some of the entrained air may be lost during the stoppage, often with detrimental effect on the flow behaviour of the material. One possible solution to this problem is to blow air into the material via a perforated piece of pipe located in the material close to the

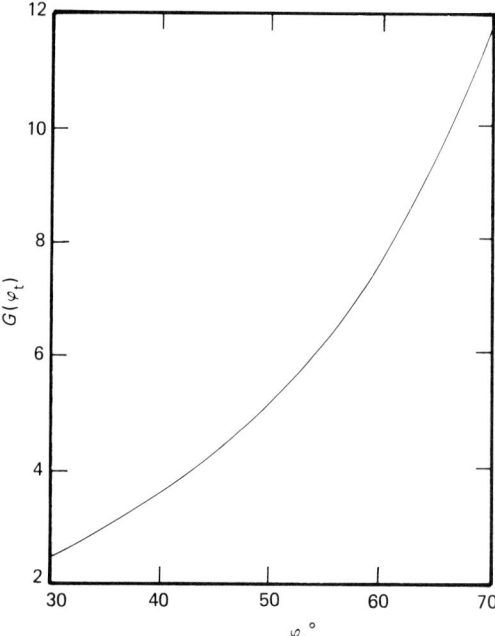

Figure 3.30 Function $G(\varphi_t)$ for no piping (funnel-flow silos) (Jenike, 1964).

outlet of the hopper; this simple device is known as an 'air lance' and its action dilates the powder in the region of the outlet, thus causing the blockage to collapse.

Air can also be introduced through the sides of the hopper; Williams *et al.* (1983) report that the best position for air entry in this method of injection is about four times the orifice diameter above the hopper opening. A modification of this type of device is to mount sintered plates on the walls of the hopper. Air injection through these porous plates will, in some cases, be sufficient to restart the flow, particularly when the bulk material is adhesive as well as cohesive.

An improvement to the mechanical air lance is the air cannon (Rappen and Wright 1983) which expels an explosive charge of compressed air directly into the hopper at the level where the blockage is expected. Rappen and Wright (1983) report a number of successful case studies involving the application of air cannons to initiate flow in bunkers of up to 100 tonnes following a period of intermission in delivery.

An alternative method of promoting bulk flow through openings is by the application of vibration. Little design information exists for such devices and if care is not exercised, vibration may actually cause further compaction, thus aggravating the problem. In such cases, the process engineer has to rely almost entirely on past experience and manufacturers' information.

Another reason for using a flow-control device is because the quantity of bulk material discharged through the opening of a mass-flow silo often exceeds the process requirement of the material. In such cases it is usually possible to use some form of mechanical feeder to restrict the discharge to the desired rate: the use of feeders is considered in more detail in Chapter 7.

References

ARNOLD, P.C. and MCLEAN, A. G. (1976a). *Powder Technol.*, **15**, 279–281
ARNOLD, P. C. and MCLEAN, A. G. (1976b). *Powder Technol.*, **13**, 255–261
ARNOLD, P. C., MCLEAN, A. G. and ROBERTS, A. W. (1980). *Bulk Solids: Storage, Flow and Handling*. TUNRA Publications
BORG, L. (1982). *Ger. Chem. Eng.*, **5**, 59–63
ECKHOFF, R. K. and LEVERSEN, G. (1973). *Powtech 73*. International Technology and Bulk Solids Exhibition and Conference, Powder Technology, UK
ENSTAD, G. (1975). *Chem. Eng. Sci.*, **30**, 1273–1283
GAYLORD, E. H. JR. and GAYLORD, C. N. (1984). *Design of Steel Bins for Storage of Bulk Solids*. Prentice-Hall, Inc
JENIKE, A. W. (1961). Bulletin No. 108, Utah Eng. Experimental Station, Salt Lake City
JENIKE, A. W. (1962). Trans. Instn. Chem. Engrs. **40**, 264
JENIKE, A. W. (1964). Bulletin No. 123. Utah Eng. Experimental Station, Salt Lake City
JENIKE, A. W., ELSEY, P. J. and WOOLEY, R. H. (1959). Bulletin No. 96. Utah Eng. Experimental Station, Salt Lake City
JENIKE, A. W. and LESER, T. (1963). 4th Int. Congress on Rheology, 125
JENIKE, A. W. and JOHANSON, J. R. (1986). Flow of solids. *Newsletter* (Jenike and Johanson) Vol. VI, No. 1, p. 3

JOHANSON, J. R. and JENIKE, A. W. (1962). Bulletin no. 116, Utah Eng. Experimental Station
JOHANSON, J. R. and COLIJN, (1964). *Iron and Steel Engineer* (October issue) 85–104
RAPPEN, A. and WRIGHT, H. (1983). *2nd Int. Conf. on Design of Silos for Strength and Flow.* Powder Advisory Centre, London, pp. 423–434
RAVENET, J. (1983). Bulk Solids Handling, March Vol. 3. No. 1
RICHARDS, J. C. (1966). *The Storage and Recovery of Particulate Solids.* I. Chem. Working Party report, Chapter 8
SAFARIAN, S. S. and HARRIS, E. C. (1984). *Design and Construction of Silos and Bunkers*, Van Nostrand Reinhold, New York
SCHUBERT, H. (1979). *Chemie Ingenieur Technik*, **51** (4), 266–277
SCHUBERT, H. (1984). *Powder Technol.*, **37**, 105–116
TAMURA, T. and HAZE, H. (1985). *Bulk Solids Handling*, **5** (3) 633–641
WALKER, D. M. (1966). *Chem. Eng. Sci.*, **21**, 975–997
WALKER, D. M. (1967). *Powder Technol.* **1**, 228
WILLIAMS, J. C. (1983). *Enlargement and Compaction of Particulate Solids*, N. G. Stanley-Wood, (ed.). Monographs in Chemical Engineering. Butterworth
WILLIAMS, J. C. and BIRKS, A. H. (1965). *Rheol. Acta*, **4** (3), 170
WILLIAMS, J. C., HEAD, J. M. and AHUMADA, J. J. (1983). *2nd Int. Conf. on Design of Silos for Strength of Flow.* Powder Advisory Centre, London, pp. 401–423
WRIGHT, H. (1970). PhD thesis, Univ. of Bradford. UK
WRIGHT, H. (1972). Trans. of ASME. Paper No. 2. 72–MH–7

Symbols

A	cross-sectional area of shear cell or bin
B	size of silo outlet for mass flow
B^*	size of silo outlet for funnel flow
C	circumference of bin
d	particle diameter
f	bulk material strength
f_c	unconfined yield strength of material
$f_{c(crit)}$	unconfined yield strength for no arching
g	gravity acceleration
h	grain height
$H(\alpha)$	function defined by Equation 3.8
i	constant = 1 for axisymmetric flow
	= 0 for plane flow
K	pressure coefficient
L	length of outlet opening
p	mean stress in the hopper outlet region
p_s	major stress acting along the surface of an arch
$\bar{p}_1$	major principal stress acting on an arch during its formation
$\bar{p}_2$	minor principal stress
$\bar{p}_R$	value of the mean stress at the transition
p_v	vertical consolidating pressure in funnel-flow silos
p_w	wall normal stress
R	radial distance from hopper vertex to transition point
r	radial distance from hopper outlet to its vertex
S_w	shear stress
S	shear stress at the wall
t	thickness dimension

Design of storage vessels for particulate storage   77

X	constant given in Equation 3.18
Y	constant given in Equation 3.19
W^*	no-arching width for funnel-flow slot silo
α	hopper half-angle
δ	effective angle of internal friction
μ	coefficient of wall friction
φ	angle of internal friction
φ_w	angle of wall friction
φ_t	static angle of internal friction

Example

It is necessary to design a mass-flow storage silo for a particular solid product of mean bulk density 960 kg/m^3. The average hold-up time in the silo is 48 hours and Table 3.3 gives the relevant time-consolidated test results from a Jenike shear cell.

Solution

The design procedure in 'Summary of design method for mass-flow silos' (p. 71) is followed:
  1. Figure 3.27 shows the family of yield loci obtained from the test results.
  2. The Table below gives the values of $\bar{p}_1$ and f_c for each yield locus:

$p(kN/m^2)$	33.8	16.2	9.0
$f_c(kN/m^2)$	10.3	5.7	3.6

  3. Figure 3.31 gives the failure function plot.
  4. Figure 3.26 gives the value of the effective angle of internal friction, which is 40 degrees.
  5. Angle of wall friction, φ_w, is 23 degrees.
  6. Equations 3.23 and 3.25 are used to evaluate the hopper half-angle, α.

Table 3.3 Time-consolidated test results (see example)

Test	1		2		3	
Normal stress $p(kN/m^2)$	Shear stress $S(kN/m^2)$	Normal stress $p(kN/m^2)$	Shear stress $S(kN/m^2)$	Normal stress $p(kN/m^2)$	Shear stress $S(kN/m^2)$
3.11	4.66	1.74	2.61	1.37	1.80
6.22	6.84	2.98	3.36	1.99	2.11
8.08	8.08	4.35	4.23	2.49	2.43
10.57	9.64	6.72	5.47	3.11	2.74
13.37	11.19			3.86	3.11

78 Design of storage vessels for particulate storage

Thus:

Opening	α (degrees)
Circular or square	23
Rectangular	30

7. Equation 3.22 is used to evaluate ff. Thus:

Opening	ff
Circular or square	1.5
Rectangular	1.6

8. Figure 3.31 shows the FF and the ff lines. The point of intersection is:

Opening	$f_c(\text{crit})$ (kN/m^2)
Circular or square	2.1
Rectangular	2.2

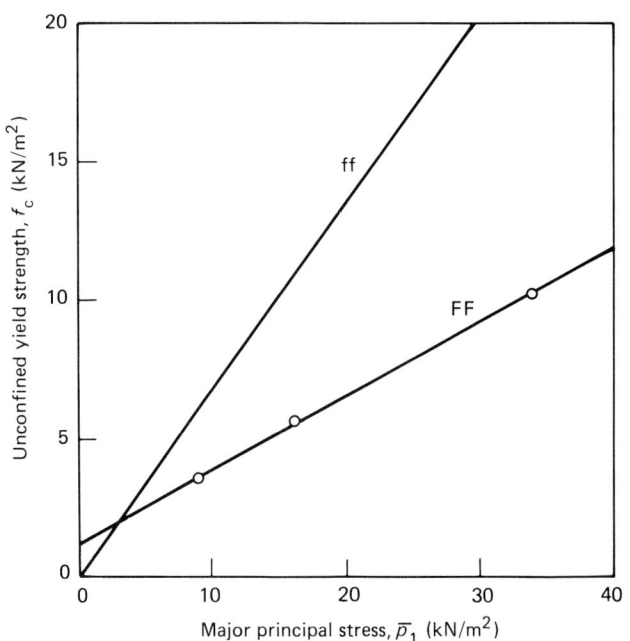

Figure 3.31 Flow factor (ff), Flow function (FF) plot.

9. The value of $H(\alpha)$ is obtained using Equation 3.8. Thus:

Opening	$H(\alpha)$
Circular or square	2.4
Rectangular	1.2

10. Minimum hopper opening, B, is evaluated using Equation 3.26. Thus:

Opening	B(m)
Circular or square	0.51
Rectangular	0.27

The following Table gives the final results:

Hopper	α (degrees)	B (m)
Circular or square	23	0.51
Rectangular	30	0.27
($L \geq 3B$)		$L = 0.81$

Chapter 4
Gravity flow of particulate solids

Introduction

In numerous commercial processing operations involving particulate solids, not only is it essential to ensure uninterrupted discharge of the material from bins and silos, but it is also imperative to control the rate of flow from such equipment.

The existing experimental information indicates conclusively that the rate of discharge of solids through apertures depends upon many factors, including:

(i) *Solids properties* — size, size distribution, shape, bulk and particle density, angles of internal and effective friction.
(ii) *Vessel and outlet characteristics* — aperture shape and size, wall roughness, presence of inserts.
(iii) *Operating conditions* — unaided (gravity) or aided (pressurized, mechanical or air-injected) flow.
(iv) *Interstitial fluid properties* — significant only at particle sizes smaller than about 500 μm.

A number of investigators have extended the general equations of motion and continuity governing the flow of fluids to cover the discharge of particulate solids from silos with varying degrees of success (Savage, 1967; Johnson, 1965; Carleton, 1972; Williams, 1977; Davidson and Nedderman, 1973; Arnold *et al.*, 1981).

The majority of workers, however, claim that such theoretical analyses are inappropriate because of the inherent differences that exist between the flow of fluids and movement of solid particles. Indeed, Stepanoff (1969) points out that unlike fluids, bulk materials cannot flow; under the action of gravity, solid particles roll, slide or fall. In the absence of any reliable theory, these researchers have fitted empirical equations to their experimental data (Franklin and Johnson, 1955; McDougall and Evans, 1966; 1969; Beverloo *et al.*, 1961; Al-Din and Gunn, 1984) or have used dimensional analysis to describe their results (Fowler and Glastonbury, 1959; Rose and Tanaka, 1959; Memon and Foster, 1985). However, a review of these publications indicates large differences in the form of the proposed correlations and in the significance of the pertinent variables.

Gravity flow of particulate solids 81

The aim of the present chapter is to clarify some of these points and to review the relevant literature pertaining to gravity flow of particulate materials from storage vessels.

It was established in Chapter 2 that during discharge from silos the pressure within the fill increases almost exponentially with height to an asymptotic value at a critical depth which is normally 2–5 bin diameters. In a deep mass-flow silo, the pressure remains constant at this value till the transition point is reached. At the level of transition, there is an abrupt change in pressure. From this point onward, the pressure decreases to almost zero at the outlet (Figure 4.1).

An important consequence of this pressure profile is that the weight of the ensiled material is mostly supported by the side walls of the container and not transmitted to the base. In other words, the head of bulk material above the outlet is not entirely available for flow. Thus, in deep silos, the rate of discharge is practically unaffected by height of grains above the orifice, a point that has been well established experimentally (Newton *et al.*, 1945; Beverloo *et al.*, 1961; Nedderman, 1985).

The bulk density of the material follows a similar trend to that of the pressure (Figure 4.1). As the bulk density changes with grain height, so does the interstitial voidage; it initially decreases as the bulk density increases and then increases as bulk density decreases. High-speed photographs (Brown and Richards, 1970) have revealed that close to the aperture the particles begin to lose contact with each other and thus fall almost freely under gravity with velocities that are about 0.1% of their terminal settling values. Moreover, experimental observations suggest that provided

$$D \geqslant 1.3B \qquad (4.1)$$

discharge is practically unaffected by container diameter (Harmens, 1963).

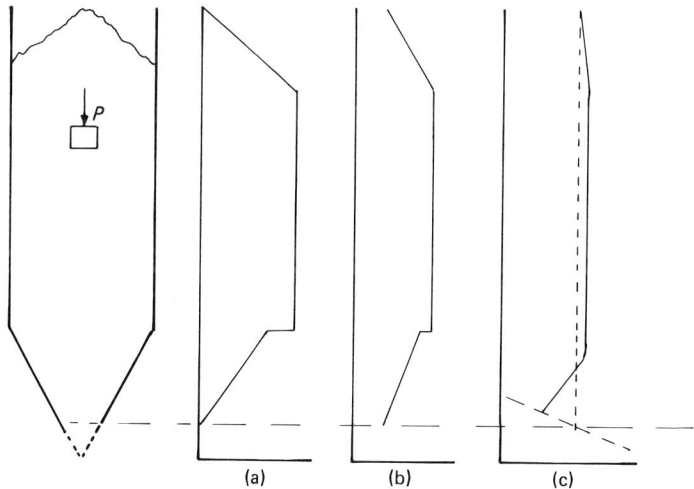

Figure 4.1 Particles pressure, bulk density and interstitial pressure profile during discharge (Bruff and Jenike, 1967–8). (a) Particle pressure profile; (b) bulk density profile; (c) interstitial fluid pressure profile.

82   Gravity flow of particulate solids

It is also interesting to note that for apertures with equal flow area, the rate of discharge depends on the shape of the opening; the rate of flow for a circular orifice is substantially higher than that from a square opening, which in turn provides a higher rate of discharge than a semicircular orifice, with a rectangular aperature giving the lowest throughput (Kotchanova, 1970; Brown and Richards 1970).

With coarse, free-flowing particles, experience suggests that uninterrupted flow occurs when: (Harmens, 1963; Newton *et al.*, 1945)

$$B > 6d_p \tag{4.2}$$

while for $B = 6d_p$ discharge becomes irregular and at $B < 4d_p$ flow is likely to stop completely due to mechanical arching at the outlet.

The rate of discharge is also influenced by the shape of the particles; spherical particles discharge faster than angular (irregular) particles (Brown and Richards, 1959).

With fine cohesive materials, it is generally believed that if mass-flow prevails in the silo, then the rate of discharge is not affected by cohesion (Nedderman, 1985). However, experimental evidence indicates that as the voidage increases towards the outlet, the interstitial fluid pressure decreases very sharply to a value below the ambient pressure immediately above the opening (Nedderman, 1983; Williams *et al.*, 1983). This negative pressure gradient in turn causes an upward flow of air which is thought to oppose the downward movement of solid particles. Although the effect of fluid drag on the motion of particles is small for coarse grains ($d_p > 600$ μm), it is very significant when fine particles are involved. As a consequence, there is a marked decrease in the rate of fine solids discharge to well below values predicted by most theoretical and empirical equations developed for coarse particles. Such observations indicate that the rate of discharge increases slowly with decrease in particle size up to a maximum value (Figure 4.2). Beyond this point, further decrease in particle size results in a corresponding decrease in the rate of discharge (Spinks and Nedderman, 1979).

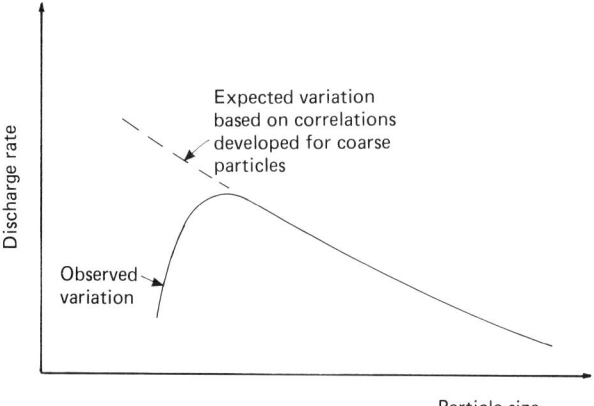

Figure 4.2 The effect of particle size on the rate of discharge from silos.

Furthermore, measurements over a long period of continuous gravity discharge (Harmens, 1963) reveal that solids draw-down through the orifice is not uniform, but is of a pulsating nature. Visual observations (Brown and Richards, 1970) have shown that a moving bed of particles forms an arch over the opening; this dynamic arch is in a constant and rapid state of formation and collapse for as long as flow continues. Moreover, in the vicinity of the aperture, flow is almost radial (Brown and Richards, 1970; Jenike 1962). As a result, in this region particle motion is directed radially inward towards a point below the aperture.

A consequence of such a radial motion is that a vena contracta is often observed (Brown and Richards, 1970; Beverloo *et al.*, 1963), i.e. there is no discharge within a short annulus round the edge of the outlet (Figure 4.3). For flat-bottom bins, the size of the annulus depends primarily upon particle properties, but is unaffected by aperture shape and size.

Thus, experimental results are often expressed in terms of the reduced orifice diameter defined as $(B - kd_p)$ instead of the diameter of the orifice B; kd_p is the thickness of the annulus with k, known as the particle shape constant, obtained experimentally. Reported values of k range between 1.0 and 3.0 for discharge through flat-bottom bins, but values well below 1.0 have been reported for k for discharge through converging channels, indicating some effect of vessel geometries upon k (Nedderman, 1985). Nedderman (1985) also points out that it is more plausible to consider the width of the 'empty space' as the thickness of the shear layer next to either the wall of the container or the stationary solid region in the silo.

From what has been said so far, it seems reasonable to divide solids discharge through orifices into two broad categories depending on whether the particles being discharged are coarse or fine.

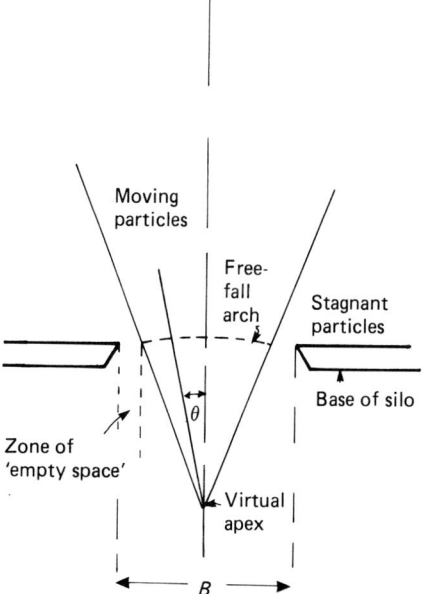

Figure 4.3 Particles discharge towards the hopper outlet and the formation of the 'empty space'.

Gravity discharge of coarse particles

Dimensional analysis and empirical equations

(i) Flat-bottom bins

For dry, coarse particles discharging from an aperture, a simple analysis based on conventional equations of mechanics for free falling bodies gives:

$$v_0 = \sqrt{2gh} \tag{4.3}$$

where g is gravity acceleration and h is the height of free fall; this is the height from which the particles must fall in order to reach a velocity v_0 at the outlet.

Substituting for the velocity in terms of the rate of mass discharge, W, material bulk density, ρ_b, and assuming circular opening with flow area, $\pi B^2/4$, yields;

$$W = \{\sqrt{2\pi}/4\}\rho_b g^{0.5} B^2 h^{0.5} \tag{4.4}$$

Dimensional analysis (Fowler and Glastonbury, 1959), on the other hand, indicates that:

$$W \propto B^{2.5} \rho_b g^{0.5} \tag{4.5}$$

From Equations 4.4 and 4.5, it may be observed that the height of free fall, h, is of the same order of magnitude as the size of the aperture. Thus, allowing for the 'empty space' round the edge of the orifice, Equation 4.4 may be rewritten as:

$$W = \lambda \{2\pi/4\} \rho_b g^{0.5} (B - kd_p)^{2.5} \tag{4.6}$$

λ is a constant which is obtained experimentally; it is analogous to the conventional discharge coefficient for fluid flow through orifices (Stepanoff, 1969).

Beverloo *et al.* (1961) proposed the following empirical correlation for solids discharge from central conical apertures:

$$W = 0.58 \, \rho_b g^{0.5} (B - kd_p)^{2.5} \tag{4.7}$$

in SI units with a maximum deviation of 10%.

Noting that the value of the term in { } in Equation 4.6 is almost unity, the value of the constant λ is therefore 0.58 and is independent of the angle of internal friction, φ.

Experimental data of Beverloo *et al.* (1961) also show that the value of the particle shape constant, k, varies in the range 1.3–2.9, depending upon surface characteristics of the bulk material; an average value of 1.4 was recommended by Beverloo *et al.*

Other researchers have also reported reasonable agreement between their experimental observations, and have predicted values for discharge rates using Beverloo's correlation (Nedderman, 1985; Murphy, 1986). Harmens (1963), Memon and Foster (1985) and Nedderman (1985), pointed out that the choice of ρ_b in Equation 4.7 is questionable, particularly since discharge is practically independent of grain head and that bulk density decreases appreciably as particles approach the aperture.

Harmens (1963) suggests that particle density is a more appropriate parameter, while Nedderman (1981; 1985) argues that a bulk density defined as the ratio of mass flow to volumetric flow rate gives the least scatter in data when used in the Beverloo correlation.

For non-circular apertures, Beverloo et al. recommended the use of an effective hydraulic diameter, B'_h, defined as:

$$B'_h = (B_h - kd_p) \tag{4.8}$$

which is then used to obtain an effective aperture area, A'_h; the proposed equation for discharge rate is:

$$W = 0.75 \, \rho_b A'_h g^{0.5} (B'_h)^{0.5} \tag{4.9}$$

Al-Din and Gunn (1984) noted that such an approach can lead to the unrealistic conclusion that for flow through a slot hopper with a large length-to-width ratio, discharge rate is practically independent of slot length. To overcome this difficulty, they proposed the following alternative equation for design calculations:

$$W = 0.406 \, F_o F_p \rho_b g^{0.5} (A/B^2)(B - d_p)^{2.5} (B/d_p)^{0.154} \tag{4.10}$$

where B is the width of the slot opening or the diameter of the circular orifice and A is the flow area. F_o and F_p are orifice and particle constants respectively; their values found experimentally by Al-Din and Gunn are given below:

F_o = 1.0 for circular and half square triangular openings
   = 1.4 for rectangular and elliptical openings.
F_p = 1.0 for round particles
   = 0.75 for sharp-edge, isometric particles
   = 0.32 for isometric particles.

Nedderman et al. (1982) state that for slot openings the following expression, which is more convenient than Equation 4.10, may be used:

$$W = 4\sqrt{2}\lambda/\pi \rho_b (L - kd_p)(B - kd_p)^{3/2} g^{0.5} (B'_h)^{0.5} \tag{4.11}$$

This equation is based on the correlation originally developed by Fowler and Glastonbury (1959) for non-circular apertures using dimensional analysis:

$$W = 0.236 A \sqrt{2gB_h} \, \rho_b \left(\frac{B_h}{d_p}\right)^{0.185} \tag{4.12}$$

Brown and Richards (1970) proposed two separate equations for mass discharge rate depending upon orifice shape. These are:

$$W = (\sqrt{2}/15)\pi \rho_b g^{0.5} (B - kd_p)^{2.5} \psi_d \tag{4.13}$$

for circular openings and

$$W = \sqrt{2/3} \, \rho_b g^{0.5} L(B - kd_p)^{3/2} \psi_s \tag{4.14}$$

for slot openings.

The values of the two empirical constants, ψ_d, and ψ_s, may be taken as unity for all practical purposes.

86   Gravity flow of particulate solids

(ii) Converging channels

With flat-bottom bins, particles are stationary close to the walls of the container, but slide and roll over each other in the central region before falling through the opening. In silos with steep converging channels, particles also slide over the hopper wall. Most researchers agree that solids movement close to the wall will also affect the rate of discharge through the opening. However, no consensus of opinion has yet emerged as to the exact form of the dependency of the rate of discharge upon hopper angle.

Deming and Mehring (1929) investigated the effect of cone angle upon the rate of solids discharge over a wide range of variables (Table 4.1) and proposed the following empirical equation:

Table 4.1 Data from Deming and Mehring (1929)

Material	Average particle size d (mm)	Bulk density (g/cm^3)	Coefficient of wall friction ($\tan \varphi_w$)	Orifice size B (mm)	Hopper half-angle	Observed mass flow rate W (g/s)
Mustard seed	2.15	0.745	0.48	10.06	15	10.35
				10.06	30	7.79
				15.09	45	20.89
Kale seed	1.724	0.688	0.416	10.06	15	11.26
				10.06	30	8.96
				15.09	45	23.91
Urea pellets	0.663	0.757	0.405	10.06	15	15.15
				10.06	30	12.17
				10.14	45	11.11
Velch seed	3.38	0.818	0.415	16	45	23.34
				19.09	45	39.12
Marbles	13.5	1.322	0.412	73.03	30	2150.54
Glass beads	3.54	1.572	0.418	10.06	15	21.56
				10.06	30	16.03
				15.09	45	43.63
Phosphate rock	0.427	1.25	0.69	5.02	15	2.98
				4.97	30	2.45
				4.98	45	2.22
Phosphate rock	0.161	1.28	0.69	2.94	15	0.96
				2.99	30	0.84
				3	45	0.81
Lead shot	1.78	6.545	0.355	10.06	15	119.90
				10.06	30	92.59
				15.09	45	245.10
Lead shot	2.03	6.595	0.355	10.06	15	110.38
				10.06	30	86.81
				15.09	45	23.15

$$W = \frac{1.67 B^{2.5} \rho_b}{\tan \varphi [34.6 + (67.4 + 444 \sin \alpha)][(d_p/B) + 0.13 - 0.161 \tan \varphi]} \quad (4.15)$$

with B and d_p in mm and W in g/s.

Franklin and Johnson (1955) pointed out that Equation 4.15 is only applicable for cone angles of up to $(180-2\varphi)$, and suggested an alternative expression for cylindrical silos. This is:

$$W = \frac{\rho_b B^{2.93}}{(6.288 \tan \varphi + 23.16)(d_p + 1.889) - 44.9} \times \frac{\cos \varphi + \cos \delta}{\cos \varphi + 1} \quad (4.16)$$

with B in in. and W in lb/min.

Rose and Tanaka (1959) used dimensional analysis to develop the following expression for discharge:

$$W = B^{2.5} \rho_b g^{0.5} (B/d_p - 3)(S - 5) \exp\{-7.7 \times 10^{-6} [c/(d_p^3 \rho_b g)]\} fn(\alpha, \theta) \quad (4.17)$$

with

$$fn(\alpha, \theta) = [(\tan \alpha \tan \theta)]^{-0.35} \quad \alpha < 90 - \theta$$
$$fn(\alpha, \theta) = 1.0 \quad \alpha > 90 - \theta$$

In Equation 4.17, S is a shape factor, θ is the included semi-angle of the flow channel and the term in { } accounts for cohesive forces between particles.

Theoretical equations

Brown and Richards (1970) assumed the motion of the particulate phase to be governed by Bernoulli's equation. They derived theoretical expressions for discharge through plane and axisymmetric flow geometries by considering the changes in the potential and kinetic energies of the moving grains and observing that the flow field immediately above the free-fall arch is purely radial. These workers assumed, with no apparent justification, that the total energy of the flowing solids decreases progressively as the particles approach the plane shown in Figure 4.3; at the outlet the total energy is zero, that is:

$$\frac{d[K \cdot E + P \cdot E]}{dR} = 0 \text{ at } R_0 = \frac{B - kd_p}{2 \sin \theta} \quad (4.18)$$

For a powder with constant bulk density their final design equations are (Brown and Richards, 1970):

$$\frac{W}{\rho_b L g^{0.5} (B - kd_p)^{3/2}} = \frac{\sqrt{\theta}}{\sqrt{2} \sin \theta} \quad (4.19)$$

for slot openings and

$$\frac{4W}{\rho_b g^{0.5} \pi (B - kd_p)^{5/2}} = \frac{2(1 - \cos^{3/2} \theta)}{3(\sin^{5/2} \theta)} \quad (4.20)$$

for circular apertures.

Equations 4.19 and 4.20 agree reasonably well with experimental observations, indicating that solids discharge rates are independent of grain height. Nedderman *et al.* (1982) point out that, this is hardly surprising as Brown and Richards only considered the kinetic and potential terms, but ignored the effect of the pressure term in their energy equation for the moving particles. This, in effect, ensures that grain height does not enter the analysis.

Nedderman *et al.* (1982) state that the 'minimum energy theorem' of Brown and Richards (Equations 4.19 and 4.20) overestimate the rate of mass discharge by a factor of 2.0. Williams *et al.* (1983) measured rates of discharge from cylindrical silos that were 10% above those predicted by Equation 4.20. It is also interesting to note that Equations 4.19 and 4.20 are of the same form as Beverloo's correlation (Equation 4.7), i.e.

$$W = fn(\theta)\rho_{\text{b}}g^{0.5}(B - kd_{\text{p}})^{2.5} \tag{4.21}$$

A difficulty in using Equations 4.19 and 4.20 is in the determination of the semi-angle of the flow channel, θ. For mass-flow silos, this may be taken as the hopper half-angle, α.

Other theoretical analyses (Davidson and Nedderman, 1973; Williams, 1977) also regard the bulk material as a continuum, but solve the Navier–Stokes equations of motion with appropriate boundary conditions. A basic assumption in the derivation of most of these equations is that the ratio of major to minor stresses, K, defined as:

$$K = \frac{1 + \sin \delta}{1 - \sin \delta} \tag{4.22}$$

is a constant throughout the fill.

For coarse, free-flowing bulk solids discharging through conical mass-flow silos, Williams (1977) solved the equations of motion along the wall and along the centre line of the hopper in order to derive lower and upper limit expressions for the rate of flow. The effects of the angle of internal and wall friction upon solids draw-down are accounted for in the theory, but to allow for the influence of particle diameter, the concept of reduced orifice diameter was employed. The upper limit for the rate of mass discharge is given by:

$$W = (\pi/3)\rho_{\text{b}}g^{0.5}(B - kd_{\text{p}})^{2.5} \sqrt{\frac{(1 + K)}{2(2K - 3)} \frac{(1 - \cos^{3/2}\alpha)}{\sin^{5/2}\alpha}} \tag{4.23}$$

Compared to Equation 4.23, the expression for the lower limit is more complicated, involving frictional properties of the powder at the wall. However, predicted discharge rates appear to be extremely insensitive to changes in wall friction. Davidson and Nedderman (1973) ignored the effect of wall friction from their analysis by assuming frictionless hopper walls. For coarse particles discharging through a mass-flow silo, these workers solved the equation of motion in the radial direction to obtain the following expression for the rate of discharge for bulk powders of industrial significance.

Gravity flow of particulate solids 89

$$W = (\pi/4) \frac{g^{0.5}\rho_b B^{5/2}}{\sin^{1/2}\alpha} \left[\frac{1 + K}{2(2K - 3)} \right]^{1/2} \quad (4.24)$$

for cylindrical silos and

$$W = \frac{\rho_b g^{0.5} L B^{3/2}}{\sin^{1/2}\alpha} \left[\frac{1 + K}{2(K - 2)} \right]^{1/2} \quad (4.25)$$

for chisel-shaped silos with width B and length L (Nedderman et al., 1982).

Equations 4.22–4.25 confirm the important experimental observation that the rate of solids discharge is unaffected by height of grain above the opening. However, Equations 4.24 and 4.25 tend to overestimate the rate of discharge by a factor of 1.5 to 2.0. Davidson and Nedderman (1973) pointed out that this is mainly due to their assumption of smooth flow along the walls of the hopper. But since the analysis by Williams (1977) shows that the mass discharge rate is very insensitive to variations in the angle of wall friction, the overprediction of discharge rate by Equations 4.24 and 4.25 cannot be explained solely on the basis of the angle of wall friction.

Carleton (1972) derived a theoretical expression for solids velocity discharging through circular apertures based on a simple force balance on a moving elemental slice of the powder (Figure 4.4). In the case of the conical hopper, if the differential slice at a radial position R from the vertex

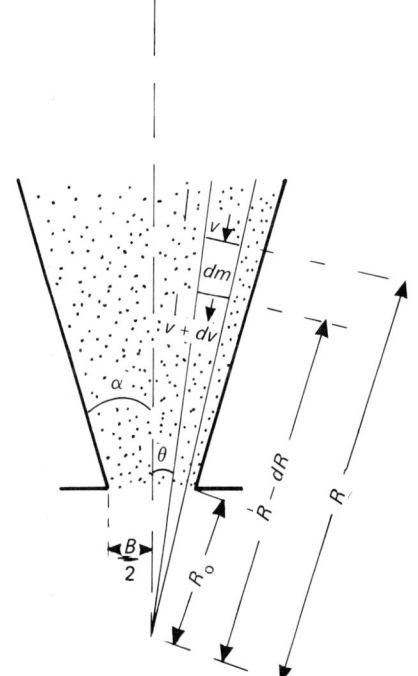

Figure 4.4 Flowing solids from hopper.

90   Gravity flow of particulate solids

has a velocity v and accelerates to a velocity $(v + dv)$ at a radial distance $(R-dR)$, then the equation of continuity gives:

$$dv/dR = 2v/R \tag{4.26}$$

and acceleration will be:

$$v \, dv/dR = 2v^2/R \tag{4.27}$$

The force on the elemental slice is given by:

$$dm \, v \, dv/dR = 2dm \, v^2/R \tag{4.28}$$

where dm is the mass of the element.

At the outlet the only force acting on the element are this inertial force (Equation 4.27) due to the vertical acceleration and the vertical component of the weight, which is $dm \, g \cos\theta$. Thus, at the outlet:

$$dm \, g \cos\theta = 2dm \, v_0^2/R_0 \tag{4.29}$$

From the geometry of the hopper $R_0 = B/2 \sin\alpha$ giving:

$$v_0 = \left[\frac{gB \cos\theta}{4 \sin\alpha}\right] \tag{4.30}$$

which is, in fact, identical to the expression developed previously by Brown and Richards (1961) using their minimum-energy theorem.

Flow rate may be obtained by integrating θ from 0 to α, but since for mass-flow silos α is normally small, it may be assumed that $\cos\theta = 1.0$ with little loss of accuracy.

Mass discharge rate is obtained by simply muliplying v_0 by the cross-sectional flow area of the orifice and bulk material density which is assumed to be constant. Thus:

$$W = \pi/8 B^{2.5}(g/\sin\alpha)^{0.5}\rho_b \tag{4.31}$$

Nedderman's kinematic model for predicting the particle velocity distribution near the outlet also assumes that solids discharge is controlled largely by conditions near the outlet with particle weight as the primary cause for motion. Experimental evidence is provided to suggest that particle motion originates at the orifice and propagates upwards as particles fall off through the opening layer by layer (Nedderman and Tuzun, 1985a,b). Nedderman and co-workers (Nedderman et al., 1983, Spinks and Nedderman, 1978; 1979; Tuzan and Nedderman, 1979) propose that near the outlet, the horizontal component of particle velocity may be expressed in terms of the vertical velocity gradient by an equation of the form:

$$u = -\xi(dv/dr) \tag{4.32}$$

For an incompressible material, Equation 4.32 may be combined with the continuity equation to give:

$$(du/dy) = \xi(d^2v/dr^2) \tag{4.33}$$

No theoretical justification is given for this expression. For flow of coarse, free flowing particles ($d_p > 1$ mm) a solution to Equation 4.33 is given by (Nedderman et al., 1982):

$$v = \frac{Q}{2\sqrt{\pi \xi y}} \exp[-(r^2/4\xi y)] \tag{4.34}$$

where the kinematic constant, ξ, is the only parameter that needs to be determined experimentally. Data by Nedderman and co-workers suggest that:

$$\xi = 2.2 \, d_p \tag{4.35}$$

However, ξ was also found to be affected by the hopper half-angle, α. Along the centre line ($r = 0$), the velocity v_c is given by:

$$v_c = \frac{Q}{2\sqrt{\pi \xi y}} \tag{4.36}$$

and hence Equation 4.34 may be rewritten as:

$$v = v_c \exp[-(r^2/4\xi y)] \tag{4.37}$$

A recent experimental investigation by Jones et al. (1985) is in basic agreement with the above kinematic model, but as Nedderman et al. (1983) point out further research is needed in this area before design recommendations can be made.

Gravity discharge of fine particles

Experimental observations by a large number of workers (Miles et al., 1968; Carleton, 1972, Bosely et al., 1969; Nedderman et al., 1983; Williams et al., 1983) reveal that with particles smaller than about 500 μm, most theoretical and empirical equations tend to overestimate the rate of discharge by as much as a factor of 10; in some cases the flow might stop completely. The interstitial fluid pressure gradient in the vicinity of the outlet is believed to be chiefly responsible for this effect.

For air-assisted (pressurized) discharge an empirical approach by Crewdson et al. (1977) suggested modifying Beverloo's correlation (Equation 4.7) by adding an extra force, due to the interstitial fluid pressure gradient at the outlet, to the weight of the material. The mass rate of discharge is then given by:

$$W = \lambda \left[g + \frac{1}{\rho_b} \frac{dp}{dR}\bigg|_{R=R_0} \right]^{1/2} \rho_b (B - k d_p)^{5/2} \tag{4.38}$$

In practice it is not always easy to measure the pressure gradient at the outlet. Nedderman et al. (1983) points out that at low rates of interstitial

92 Gravity flow of particulate solids

fluid flow ($Re_f < 10$) and provided $R \gg R_0$, then the Carman–Kozeny equation may be used to give:

$$(dp/dR)_{R=R_0} = \Delta p/R_0 \qquad (4.39)$$

where Δp is the overall pressure drop.

For high interstitial fluid flow ($Re_f > 500$) the Ergun equation is more appropriate. This gives;

$$(dp/dR)_{R=R_0} = 3\Delta p/R \qquad (4.40)$$

Thus Nedderman suggests that at high and low interstitial fluid Reynolds number, Equation 4.38 predicts that:

$$W \propto \left[g + \frac{C}{\rho_b} \times \frac{\Delta p}{R_0} \right]^{0.5} \qquad (4.41)$$

McDougall and Evans (1966; 1969) correlated their data for normal gravity and pressurized discharge from flat-bottom bins in terms of the pressure drop across the bed. Their analysis, which involves the application of Bernoulli's equation between a point below and a point above the orifice, gives the following expression for the rate of discharge;

$$W = 0.123(\Delta p + \Delta p_0)^{0.5} B^{2.5} \qquad (4.42)$$

Resnick (1972) noted that when $\Delta p = 0$, i.e. for normal gravity discharge, then $\Delta p_0 \propto g\rho_b$ and thus Equation 4.42 may be restated as:

$$W \propto \rho_b (g + K_1 \Delta p)^{0.5} B^{5/2} \qquad (4.43)$$

Reported experimental data (Nedderman, 1983) have confirmed that the form of Equation 4.41 is basically correct for conical hoppers, but the relationship between the pressure gradient at the orifice and the overall pressure drop across the fill is more complex for non-circular geometries (Nedderman et al., 1983).

Carleton (1972) stated that the pressure gradient in the orifice region of the hopper may be related to particle velocity in terms of a drag coefficient, C_d. Carleton incorporated the resulting drag force into the force balance for the moving elemental slice of the bulk material (Figure 4.4) to give:

$$\frac{2dm \; v_0^2}{R_0} = dm \; g \cos\theta - C_d \rho_b v_0^2 (d_p^2 \pi/4) \qquad (4.44)$$

where the last term in Equation 4.44 is the drag force on the particles. Noting that for most mass-flow silos, the hopper half angle is small and therefore $\cos\theta$ may be taken as unity and letting $dm = \rho_s \pi d_p^3/6$, Equation 4.44 reduces to:

$$\frac{2v_0^2}{R_0} = g - 3/2 C_d (\rho_f/\rho_s) v_0^2/d_p \qquad (4.45)$$

For spherical particles falling unhindered through the stationary fluid (air) Carleton recommends:

$$C_d = 10/Re_p^{2/3} \qquad (4.46)$$

for $2 < Re_p < 200$

Substituting into Equation 4.44 and replacing R_0 by $B/2 \sin \alpha$ gives:

$$\frac{4v_0^2 \sin \alpha}{B} + \frac{15\rho_f^{1/3}\mu_f^{2/3}v_0^{4/3}}{\rho_s d_p^{5/3}} = g \tag{4.47}$$

which is a cubic equation in $v_0^{2/3}$.

In cases where the fluid drag term is sufficiently large, e.g. with small, light particles in a dense fluid, the inertia term may be neglected and Equation 4.47 becomes:

$$v_0 = \frac{g^{3/4}\rho_s^{3/4}d_p^{5/4}}{7.6\,\rho_f^{1/4}\mu_f^{1/2}} \tag{4.48}$$

The flow rate may be obtained by multiplying v_0 by the flow area. For plane-flow geometry a similar approach yields:

$$\frac{2v_0^2 \sin \alpha}{B} + \frac{15\rho_f^{1/3}v_0^{4/3}\mu_f^{2/3}}{d_p^{5/3}} = g \tag{4.49}$$

and

$$W = \rho_b B L v_0 \tag{4.50}$$

Experimental data reported by Miles et al. (1968) and Deming and Mehring (1929) have been used to check the validity of some of the theoretical and empirical equations presented in this chapter. The results shown in Figures 4.5–4.10 indicate that for large particles Williams' equation for the upper limit and Brown and Richards' minimum-energy theorem give equally good results, while the approach by Davidson and Nedderman (Equation 4.24) overpredicted the rate of discharge considerably (values were well outside the range in this analysis). For fine particle systems, on the other hand, the only reasonable approach seems to be that of Carleton (Figure 4.11), but even with this technique predictions do not agree with experimental observations for large aperture sizes.

To increase the rate of discharge of fine particles, the most commonly adopted technique is to inject air into the region just above the orifice. In practice this is usually done by pumping the air either through a perforated tube placed centrally within the fill or through a sintered material positioned in the walls around the orifice. Detailed experimental observations suggest that solids discharge rate does not increase monotonically with increasing air flow rate. At high air-flow rates, there is a fall-off in solids discharge due to particle fluidization (Yates, 1983). Moreover, as the air-flow rate increases, the issuing jet of solids becomes more and more unstable, a phenomenon that is poorly understood at the moment (Papazoglou and Pyle, 1970; Darton, 1976).

An elegant and yet very simple alternative solution to the problem of fine particles discharge is to modify the geometry of the vessel so that the partial vacuum occurs not above the outlet region, but below it. This is achieved by attaching a short piece of standpipe to the orifice (Figure 4.12). The flowing solid particles entrain the surrounding air with them as they travel down the standpipe causing a negative pressure gradient below the aperture. Yuasa and Kuno (1972) reported experimental flow rates through circular openings in flat-bottom bins with and without a standpipe.

94   Gravity flow of particulate solids

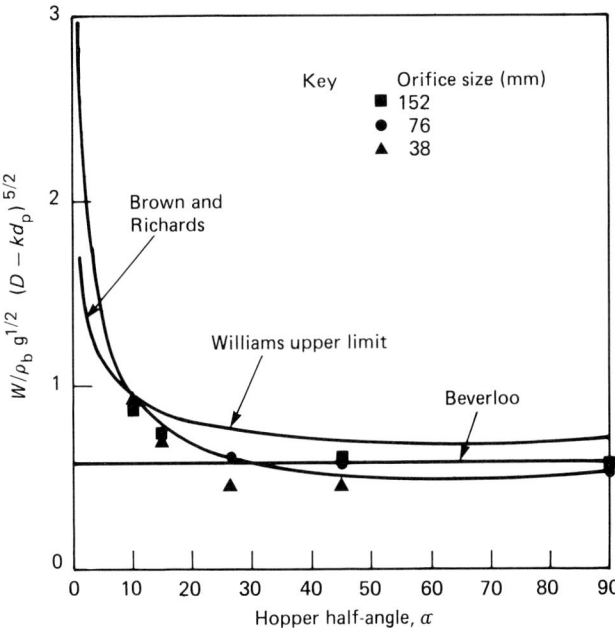

Figure 4.5 Variation of mass discharge rate with hopper half-angle. Data from Miles *et al.* (1968) for 2mm diameter gravel. Other material properties: $\rho_b = 1550$ kg/m^3, $\varphi = 40°$.

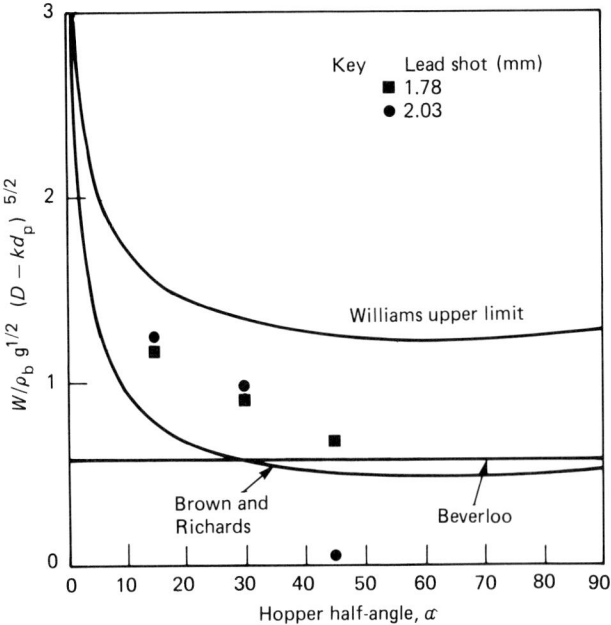

Figure 4.6 Variation of mass discharge rate with hopper half-angle. Data from Deming and Mehring (1928). See Table 4.1.

Gravity flow of particulate solids   95

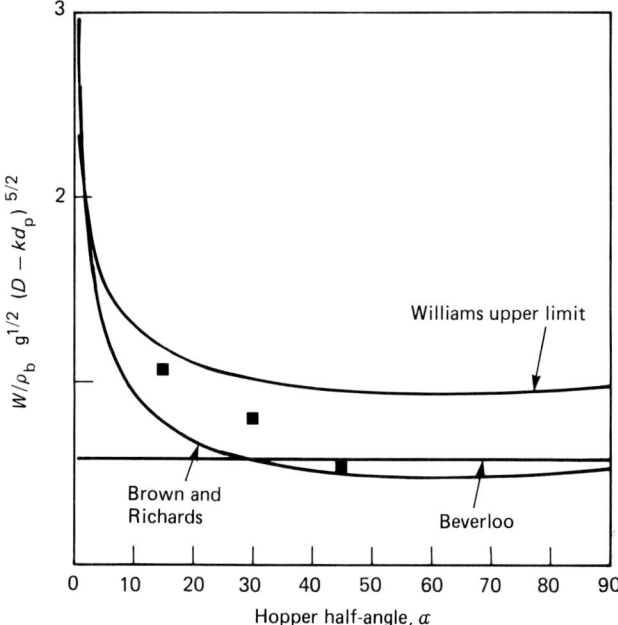

Figure 4.7 Variation of mass discharge rate with hopper half-angle. Data from Deming and Mehring (1929) for Mustard seed. See Table 4.1.

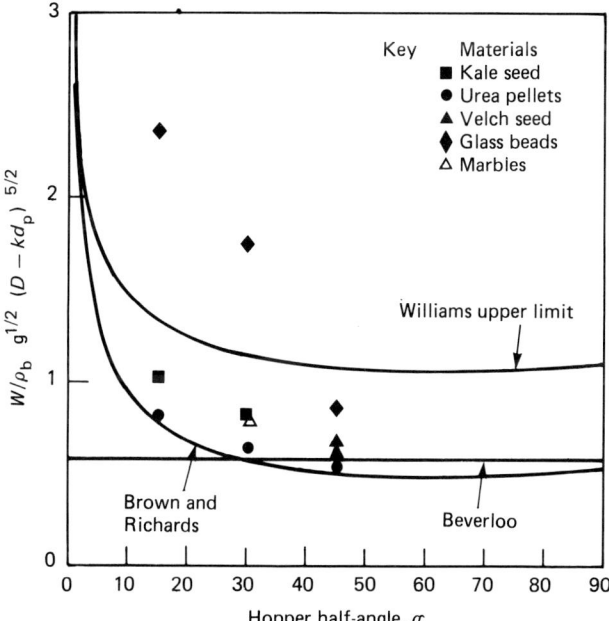

Figure 4.8 Variation of mass discharge rate with hopper half-angle for various materials. Data from Deming and Mehring (1929). See Table 4.1 for details on materials properties.

96   Gravity flow of particulate solids

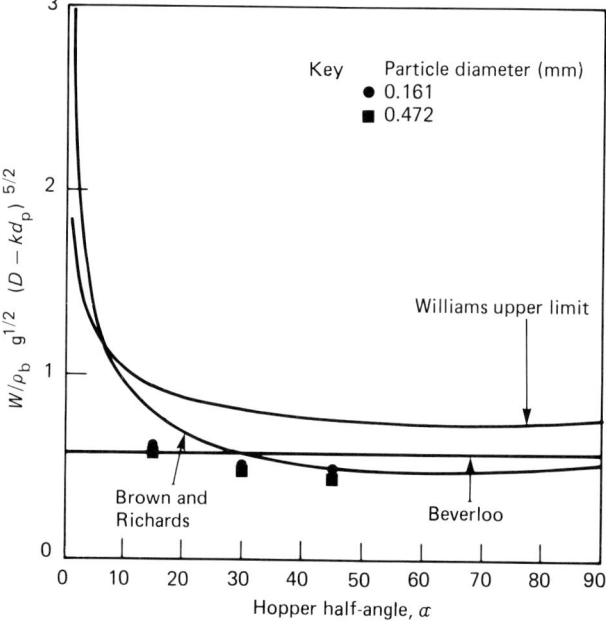

Figure 4.9 Variation of mass discharge rate with hopper half-angle for phosphate rock. Data from Deming and Mehring (1929). See Table 4.1 for other material properties.

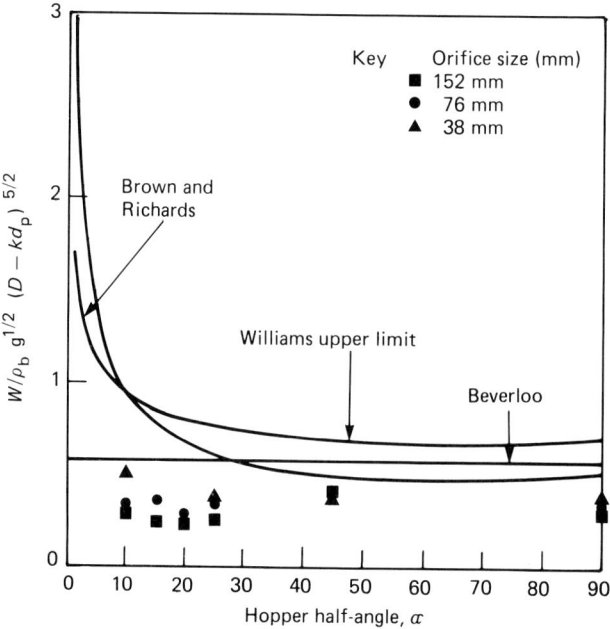

Figure 4.10 Variation of mass discharge rate with hopper half-angle for 66 μm diameter sand. Data from Miles *et al.* (1968).

Gravity flow of particulate solids   97

Their results indicate that the addition of the standpipe increases the discharge rate by as much as a factor of six. Similar results have been observed by other workers (De Jong and Hoelen, 1975; Brown and Richards, 1970).

These experimental observations also reveal that the increase in the rate of discharge through the standpipe is most pronounced in the case of fine powders, as would be expected. But a more striking feature of these investigations is that they indicate an increasing rate of efflux with increase in tube length. Yuasa and Kuno (1972) correlated their data with a simple power–law equation of the form:

$$\Delta p = K_2((WL/d_p)B^2)^{K_3} \tag{4.51}$$

where K_2 and K_3 are empirical constants; for glass beads used by Yuasa and Kuno, $K_2 = -0.00126$ and $K_3 = 0.7$.

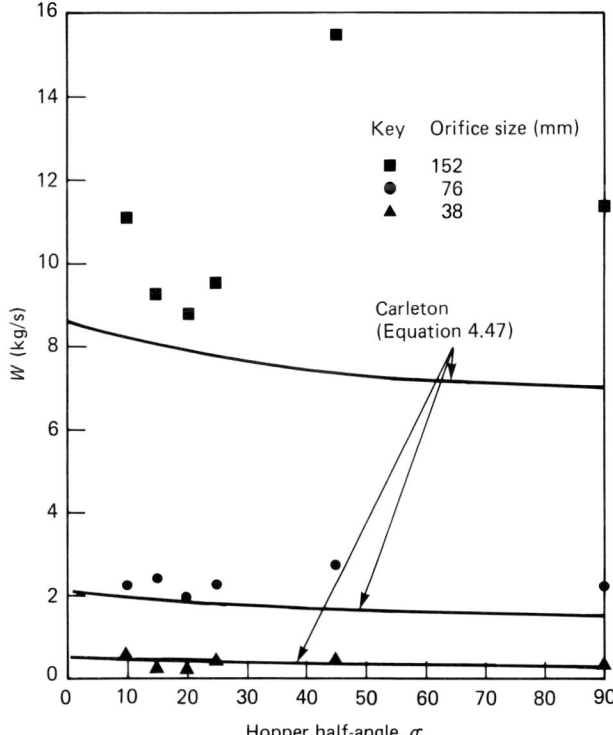

Figure 4.11 Variation of mass discharge rate with hopper half-angle for 66 μm diameter sand. Data from Miles et al. (1968). Material properties: $\rho_s = 2453$ kg/m^3 (estimated); $\rho_b = 1330$ kg/m^3, $\mu_f = 1.8 \times 10^{-5}$ kg/ms; $\rho_f = 1.0$ kg/m^3.

98 Gravity flow of particulate solids

Yuasa and Kuno (1972) also point out that a practical method to control the rate of solids discharge from the silo is by careful injection of air into the top of the standpipe.

All the equations presented so far have been derived for centrally positioned apertures. Many industrial silos are equipped with eccentric outlets. Kotchanova (1970) presented some experimental data for such systems. His results indicate that for the same orifice diameter and shape, discharge rate is higher for the eccentric outlet. However, despite their industrial significance, there has been little systematic work on flow from such openings.

Moreover, most of the research to date on solids discharge from apertures has been confined to mono-sized or narrowly classified mixtures of particles. Memon and Foster (1985) studied discharge rates of binary mixtures of large and small particles through hoppers. They reported that the presence of large particles reduces the rate of flow through the hopper

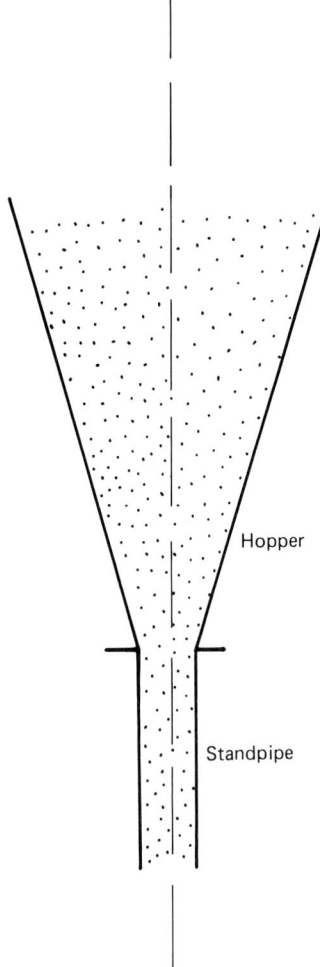

Figure 4.12 Hopper with standpipe.

but the effect depended more on the mass fraction of the large particles and less on their sizes. Clearly more work is needed on the flow of mixtures of solids through bins and silos before any conclusions can be drawn.

In practice, the silo is often equipped with internal baffles and inserts in order to achieve certain desirable flow conditions during operation. For example, in the case of funnel-flow bins, it may be necessary to eliminate dead zones in the vessel. Flow-correcting inserts have been shown to alter the flow patterns and particles velocity distribution enormously within the silo (Johnson 1966; Nedderman and Tuzun, 1985a,b). Alternatively, where excessive wall loads are expected to occur during discharge, it may be necessary to reduce the overpressure by the use of anti-dynamic tubes inserted centrally within the hopper (Ooms and Roberts, 1985).

The presence of inserts of any kind within the fill will also affect throughput from the silo. But despite their industrial significance, no reliable method exists at the present time for predicting the rate of discharge from such systems.

References

AL-DIN, N. and GUNN, D. D. (1984). *Chem. Eng. Sci.*, **39**, 121–127
ARNOLD, P. C., MCLEAN, A. G. and ROBERTS, A. W. (1981). *Bulk Solids Handling*, **1** (1), 13–23
BEVERLOO, W. A., LENIGER, H. A. and VAN DE VELDE (1961). *Chem. Eng. Sci.*, **15**, 260–269
BROWN, R. L. and RICHARDS, J. C. (1959). *Trans. Instn. Chem. Engrs.*, **37**, 108–119
BROWN, R. L. and RICHARDS, J. C. (1960). *Trans. Inst. Chem. Engrs.*, **38**, 243–256
BROWN, R. L. and RICHARDS, J. C. (1961). *Nature*, **191**, 458
BROWN, R. L. and RICHARDS, J. C. (1970). *Principles of Powder Mechanics*. Pergamon Press, New York
BRUFF, W. and JENIKE, A. W. (1967–68). *Powder Technol.*, **1**, 252–256
CARLETON, A. J. (1972). *Powder Technol.* **6**, 91–96
CREWDSON, B. J., ORMOND, A. L. and NEDDERMAN, R. M. (1977). *Powder Technol.*, **16**, 197–207
DARTON, R. C. (1976). *Powder Technol.*, **13**, 241–250
DAVIDSON, J. F. and NEDDERMAN, R. M. (1973). *Trans. Instn. Chem. Engrs.*, **51**, 29–35
DE JONG, J. A. H. and HOELEN, Q. E. J. J. M. (1975). *Powder Technol.*, **12**, 201–208
DEMING, W. E. and MEHRING, A. L. (1929). *Industr. Eng. Chem.*, **21**, 661–665
FOWLER, R. T. and GLASTONBURY, J. R. (1959). *Chem. Eng. Sci.*, **10**, 150–156
FRANKLIN, F. C. and JOHNSON, L. N. (1955). *Chem. Eng. Sci.*, **4**, 119–129
HARMENS, A. (1963). *Chem. Eng. Sci.*, **18**, 297–306
JENIKE, A. W. (1962). *Trans. Instn. Chem. Engrs.*, **40**, 264–371
JOHANSON, J. R. (1965). *Trans. Min. Engrs., AIME*, **232**, 69–80
JOHANSON, J. R. (1966). *Trans. ASME*, 224–230
JONES, A. G., ROWE, P. N. and GRAVESTOCK, N. (1985). *Powder Technol.*, **45**, 83–86
KOTCHANOVA, I. I. (1970). *Powder Technol.*, **4**, 32–37
MCDOUGALL, I. R. and EVANS, A. C. (1966). *Trans. Instn. Chem. Engrs.*, **44**, T15–T27
MCDOUGALL, I. R. and EVANS, A. C. (1969). *Trans Instn. Chem. Engrs.*, **47**, T73–T79
MEMON, M. A. and FOSTER, P. J. (1985). *PowTech.*, I. Chem. E, Symp. Series No. 91, Birmingham, UK
MILES, J. E. P., SCHOFIELD, C. and VALENTINE, H. H. (1968). *PowTech* I. Chem. E. Symp. Series No. 29, London, UK
MURPHY, J. (1986). Final-year B.Sc.(Eng.) Research Project Report. University College London, UK
NEDDERMAN, R. M. (1981). *Bulk Solids Handling*, **1** (1), 25–30
NEDDERMAN, R. M. (1983). *2nd Int. Conf. on Design of Silos for Strength and Flow*, Stratford-upon-Avon, UK

NEDDERMAN, R. M. (1985). *PowTech.*, I. Chem. E, Symp. Series No. 91, Birmingham, UK
NEDDERMAN, R. M. and TUZUN, U. (1985a). *Chem. Eng. Sci.*, **40**. No. 3, pp. 325–336
NEDDERMAN, R. M. and TUZUN, U. (1985b). *Chem. Eng. Sci.*, **40**, 337–351
NEDDERMAN, R. M., TUZUN, U., SAVAGE, S. B. and HOULSBY, G. T. (1982). *Chem. Eng. Sci.*, **37** (11), 1597–1609
NEDDERMAN, R. M., TUZUN, U. and THORPE, R. B. (1983). *Powder Technol.*, **35**, 69–81
NEWTON, R. H., DUNHAM, G. S. and SIMPSON, T. P. (1945). *Trans. Amer. Chem. Engrs.*, **41**, 215
OOMS, M. and ROBERTS, A. W. (1985). *Bulk Solids Handling*, **5** (5), 1009–1016
PAPAZOGLOU, C. S. and PYLE, D. L. (1970). *Powder Technol.*, **4**, 9–18
RESNICK, W. (1972). *Trans. Instn. Chem. Engrs.*, **50**, 289
ROSE, H. F. and TANAKA, T. (1959). *The Engineer*, **465**, 208
SAVAGE, S. B. (1967). *Int. J. Mech. Sci.*, **9**, 651
SPINKS, C. D. and NEDDERMAN, R. M. (1978). *Powder Technol.*, **21**, 245–261
SPINKS, C. D. and NEDDERMAN, R. M. (1979). *Powder Technol.*, **22**, 243–253
STEPANOFF, A. J. (1969). *Gravity Flow of Bulk Solids and Transportation of Solids in Suspension*. John Wiley, New York
TUZUN, U. and NEDDERMAN, R. M. (1979). *Powder Technol.*, **24**, 257–266
WILLIAMS, J. C. (1977). *Chem. Eng. Sci.*, **32**, 247–255
WILLIAMS, J. C., HEAD, J. M. and AHUMADA, J. J. (1983). *2nd Int. Conf. on Design of Silos for Strength and Flow*, Stratford-upon-Avon, UK
YATES, J. G. (1983). *Fundamentals of Fluidized-bed Chemical Processes*. Butterworths, London
YUASA, Y. and KUNO, H. (1972). *Powder Technol.*, **6**, 97–103

Symbols

A	flow area
A_h	hydraulic flow area
A_h'	effective hydraulic flow area
B	diameter of a circular orifice or width of a slot opening
B_h	hydraulic diameter (flow area/perimeter)
B_h'	effective hydraulic diameter (Equation 4.8)
C	constant in Equation 4.40
C_d	particles drag coefficient
D	silo diameter
d_p	mean particle diameter
dm	mass of elemental slice
F_0	orifice constant (Equation 4.10)
F_p	particle constant (Equation 4.10)
g	acceleration due to gravity
h	height of free arch at the outlet
K	ratio of major to minor principal stresses
K_1	constant in Equation 4.43
$\left.\begin{array}{c}K_2\\K_3\end{array}\right\}$	constants in Equation 4.51
k	particle shape constant
L	length of slot opening
Δp	pressure-drop across the bed of particles in the silo under pressurized flow
Δp	pressure-drop across the bed of particles in the silo under normal gravity flow

Q	volumetric discharge rate
r	radial distance from centre line
R	radial distance measured from apex
R_0	radial distance from apex to hopper outlet point
Re	insterstitial fluid Reynolds number
Re_p	particle Reynolds number
S	shape factor in Equation 4.17
u	horizontal velocity component of particles
v	vertical velocity component of particles
v_0	particles velocity at the outlet
v_c	centre line vertical velocity
y	vertical distance from base of vessel
W	mass flow rate of solids
α	hopper half-angle
δ	effective angle of internal friction
λ	constant in Equation 4.6
μ_f	interstitial fluid viscosity
ξ	kinematic constant
θ	semi-angle of flow channel
ρ_f	interstitial fluid density
ρ_b	bulk material density
ρ_s	particle density
φ	angle of internal friction
φ_w	angle of wall friction
ψ_d	constant in Equation 4.13
ψ_s	constant in Equation 4.14

Chapter 5
Pneumatic conveying of bulk solids

Introduction

The term 'flow' when applied to the transport of granular materials is a misnomer because strictly speaking only fluids can flow; bulk solids slide, roll or fall under the action of gravity (Stepanoff, 1969). The only way that bulk solids can be pumped from one location to another is when the particles are suspended in either a flowing gas (pneumatic conveying) or a flowing liquid (hydraulic transport).

Suspension of solids in a flowing gas is the subject of this chapter. The importance of pneumatic conveying of particulate solids is well recognized and widely practised in the process industry; examples of bulk materials that have been successfully transported pneumatically include granular chemicals, food and pharmaceuticals, soap powder, paper pulp, fertilizers, metal ores, cement, coal and nuclear fuel particles. The available evidence also indicates that the range of materials that can be conveyed by this method is increasing continuously as more efficient conveying systems capable of handling larger output and longer conveying distances are developed.

Some of the main advantages and disadvantages of pneumatic transportation of bulk solids are outlined in Table 5.1.

The basic components of any pneumatic conveying system are:

(i) The transport pipeline.

(ii) The gas mover. A large range of equipment is available for this purpose including:
    (a) Roots blowers and exhausters;
    (b) centrifugal fans;
    (c) screw or reciprocating compressors.

(iii) The product feeder. Any of the following may be used for this duty:
    (a) simple suction probes;
    (b) rotary feeders;
    (c) venturi feeders;
    (d) blow tanks;
    (e) standpipes.

(iv) Product/gas separator. This could be a:
   (a) cyclone;
   (b) filter;
   (c) settling chamber.
(v) Control and instrumentation.

Figure 5.1 shows a typical arrangement used in industry. In general, however, there are numerous ways in which the basic equipment can be put together to give pneumatic transport systems capable of meeting design requirements for almost any given conveying duty (Huggett and Godfrey, 1977; Kraus, 1980).

At present there is no established means of classification of pneumatic conveying systems. Essentially they can be divided into two broad groups,

Table 5.1 Advantages and disadvantages of pneumatic conveying

Advantages	Disadvantages

1. Total enclosure of conveying product. Thus:
  (i) there is little risk of mechanical accidents;
  (ii) product will remain clean and free from contamination; this is particularly suited to foodstuffs and biological materials;
  (iii) there is less chance of fire and explosion
2. Changes in flow direction can be accommodated easily
3. Saving in space is possible as the conveying line can be above ground level
4. Automation is relatively easy
	1. Particles should be dry and free-flowing
	2. Excessive particle breakage could occur during conveying of friable materials
	3. For materials that oxidize easily, an inert gas should be used as the carrier fluid
	4. With abrasive materials excessive wear and tear might result in pipes, fans and other items of equipment coming into contact with the product

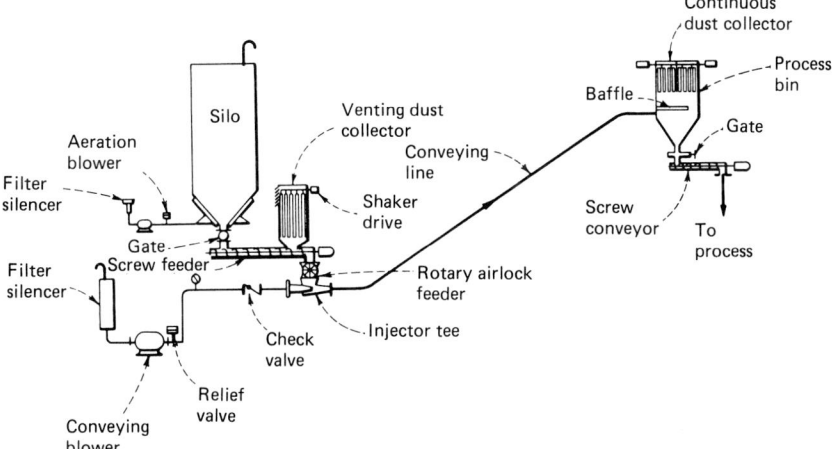

Figure 5.1 Positive-pressure pneumatic conveying system.

positive-pressure and negative-pressure transport systems, depending on whether the fluid pressure in the conveying line is above or below atmospheric. With negative-pressure systems, solids can be fed into the conveying line from several ports and discharged at one point. With positive-pressure, on the other hand, material fed from one point can be discharged to several locations. In practice, combined negative/positive-pressure systems are also possible; such conveyors are used in, for example, situations where the product needs to be collected from several locations and delivered to a number of ports.

Mason (1977) has classified solids pneumatic transport systems into pure pneumatic, mechanical-pneumatic and air-slide conveyors, depending on the method used to introduce the solid particles into the fluid stream. In pure pneumatic systems the bulk material is fed into the conveying line by the action of the carrier gas which can be either positive or negative. Mechanical-pneumatic conveyors operate in the same way except that the particles are fed into the flowline mechanically. In air-slide systems, the bulk material is fluidized in order to facilitate the movement of particles downward along the flowline. Fluidization of particles is achieved by injecting air through the carrier pipe, which is suitably sintered for this purpose.

For a given bulk solids and a specified conveying rate, the selection of the appropriate system and its design and operation is affected critically by many factors, including:

(i) Carrier gas flow rate and hence fluid turbulence and flow regime;
(ii) Solids/fluid mass ratio;
(iii) Particle properties e.g. size, size distribution, shape, density, fluidizability, hardness, cohesion and chemical nature;
(iv) Pipe size, length and roughness;
(v) Flow direction, that is, vertical, inclined or horizontal;
(vi) Moisture content.

In general, negative-pressure conveying is confined to short distances because of the limited degree of vacuum that can be attained in practice. As a result, this type of transportation has received little systematic attention in the literature and the design of such conveyors is still more of an art than a science. For that reason, negative-pressure conveying will not be considered further in this chapter.

Positive-pressure pneumatic conveying systems

Figure 5.1 shows a positive-pressure system. Essentially, the material to be conveyed is fed from a storage container into a feeder which, in turn, discharges into the gas stream flowing at a high velocity in the conveying pipeline. At the receiving end, the material is discharged into a second storage vessel where the gas is vented either directly into the atmosphere or into a filter if the material is dust laden.

With low-pressure systems, that is pressures of up to 15 p.s.i.g., rotary positive blowers and airlock valves or double-door discharge gates may be used as the air mover and the solids feeders respectively. The choice of feeder depends largely upon the pressure, temperature and material type: for example, while the rotary airlock valve is used for both negative- and positive-pressure systems, the double-door discharge gate is more suitable for high-temperature, abrasive or easily degradable materials (Kraus, 1980).

For air pressures in the range 15–45 p.s.i.g., the so-called 'medium-pressure systems', and the high-pressure units with air pressures between 45 and 125 p.s.i.g., fluid–solid pumps, e.g. screw pump conveyors, and blow-tanks are used.

The low-pressure system is fairly flexible in operation and is generally suited for conveying dry, free-flowing granular materials. It is, however, limited to conveying distances of no more than a few hundred metres. With properly designed high-pressure units, on the other hand, conveying distances of up to 3 km may be obtained (Stoess, 1983). For the detailed hardware design of the individual components of equipment of pneumatic conveying systems, the reader is referred to the excellent publications by Kraus (1980) and Stoess (1983).

(i) Horizontal transport

Consider the suspension of solid particles in a flowing gas in a horizontal pipe. At low solids–fluid mass ratios the grains are uniformly suspended with little variation in solids concentration in any direction. This is often called the 'homogeneous flow zone'. As the mass ratio is increased, the larger particles segregate towards the bottom of the conveyor pipeline; in this region, the particles roll over each other and slide along the wall of the pipe. This phenomenon is known as 'saltation' and the corresponding carrier gas velocity is referred to as the 'saltation velocity'. From this point onward, the degree of saltation increases as the solids–gas mass ratio is increased further. Likewise, as the solids–gas mass ratio increases, the overall pressure drop increases, progressively at first, then becoming erratic as flow becomes more and more unstable.

From the above simple description of flow it is evident that no single correlation can be developed to predict the overall pressure and the consequential power requirement in the gas mover for the full range of operating conditions. This is demonstrated by referring to the plot of pressure drop against velocity depicted in Figure 5.2.

In Figure 5.2, line 1 represents frictional loss for the carrier fluid alone flowing through the pipeline. Keeping the gas velocity constant, the pressure drop increases (lines 2 and 3) as solid particles are introduced into the gas stream.

Now if solids loading is maintained at a constant rate but the fluid velocity is decreased gradually, the pressure drop decreases from a value at point A to that at point B. Saltation occurs at point B as particles segregate towards the bottom of the pipe. As a result, the cross-sectional area available for flow decreases. This in turn causes an increase in the carrier gas velocity and a consequential increase in the pressure drop to a new value C. At point C a new steady state is established as the rate of solids deposition is balanced by the rate at which particles are picked and transported into the gas stream where they are suspended uniformly and conveyed along to the exit port.

The layer of particles settled at the base of the pipe, on the other hand, might remain stagnant, slide along the wall as a whole or form into small dunes in which grains are transported from one dune to another in the general direction of flow.

If at point C the velocity of the carrier gas is reduced further, the height of the settled bed of grains increases and, as a result, pressure drop increases to a new value.

(ii) Vertical transport

Leung and Wiles (1976) and Leung *et al.* (1971) have identified several types of flow in vertical pneumatic transport. The main types are:

(i) *Dilute-phase flow.* With low solids concentration ($C_v \leq 5\%$), the particles are suspended fairly uniformly and are carried upward in the flowing gas stream.

(ii) *Dense-phase flow.* With coarse and/or heavy particles transported in small-diameter pipes, slugging occurs, that is, the material is carried up by air slugs: the flow is referred to as slugging dense-phase flow. With small and light particles, particularly when flowing in large-diameter pipes, slugging does not occur and the particles are transported up in the pipeline with a large amount of internal recirculation: the flow is referred to as dense-phase flow without slugging.

(iii) *Moving-bed flow.* With this type of flow there is very little relative motion between the particles and the bulk material is transported up as a packed column.

Pressure drop–velocity plot for vertical pneumatic transport of solids is shown schematically in Figure 5.3. Line 1 once again represents the frictional loss for the flow of the carrier gas alone. With fluid velocity fixed,

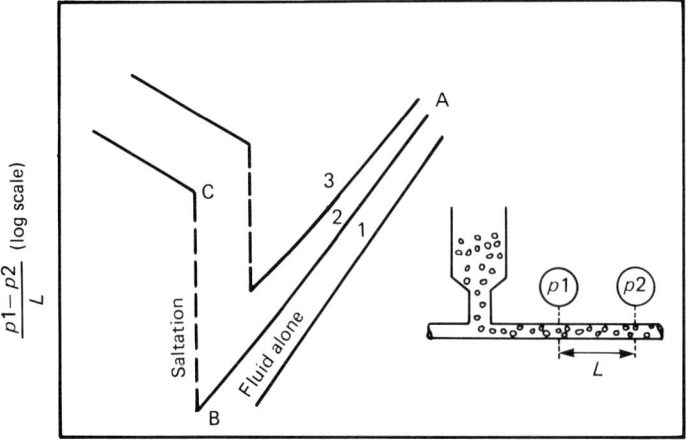

Figure 5.2 Pressure drop/flow characteristics of horizontal pneumatic conveying.

addition of solid particles into the gas stream causes the pressure to rise steadily (lines 2 and 3) provided, of course, that the gas velocity is high enough to ensure particles remaining in suspension.

With particle loading fixed, pressure drop decreases as fluid velocity is reduced. At a critical gas velocity, a minimum pressure drop is attained (point B); this drop in pressure is due largely to a decrease in gas friction in the flowline. However, as the fluid velocity is reduced further beyond point B, the pressure drop rises gradually to point C, due to the increased solids content. At point C the fluid is no longer capable of carrying the weight of the solids; a condition known as 'choking' occurs at this point, in which solid particles are usually transported by slug flow; this is analogous to slugging in fluidized beds (Yang, 1984; Leung and Wiles, 1976). With fine particles a high degree of internal recirculation occurs without slugging. This type of flow is analogous to a recirculating fluidized bed or a fast-fluidized bed and is known as 'flow without slugging' (Leung and Wiles, 1976; Yang, 1984).

Choking in vertical flow lines is, in many ways, similar to saltation in horizontal conveyors, but, in view of the existing contradictions between different investigators, it is difficult to see any relationship between the two velocities (Dalla Valle, 1942; Whetton and Broadhurst, 1952; Bragg and Kwan, 1978).

It is also interesting to note the similarities between vertical and horizontal pressure drop–velocity plots (Figures 5.2 and 5.3). Although the change in pressure in horizontal pipelines at the saltation point is more abrupt compared to flow in vertical pipes, in both cases the zone to the right of the minimum in the $\Delta p - v$ curve represents steady flow/pressure and relatively low solids–gas mass ratios, while the region to the left of this point represents high Δp and high loading rates. This difference has been used as the basis for classifying pneumatic transport systems; the region to the right of the minimum Δp represents dilute-phase transport. From a

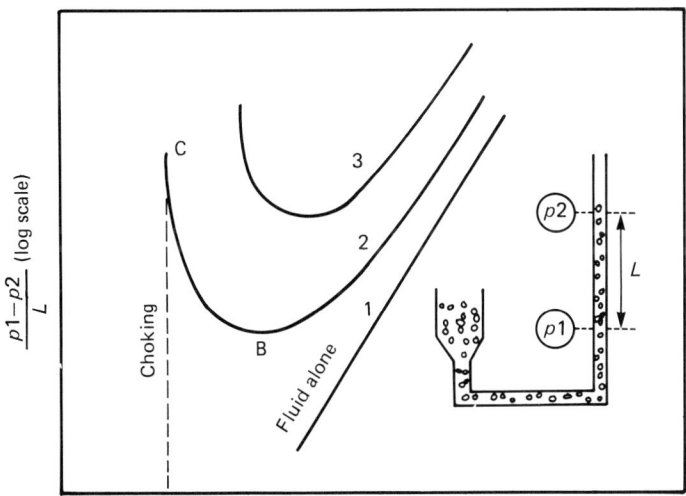

Figure 5.3 Pressure drop/flow characteristics of vertical pneumatic conveying.

design point of view it is important to ensure that solids conveying is carried out in the dilute phase as this would minimize pressure drops and hence energy requirement, pipe erosion and particle breakage during operation. Consequently, considerable work has been reported for both vertical and horizontal pneumatic conveying in this regime. This is considered below.

Dilute-phase pneumatic conveying systems

With solids concentration less than about 3–5 kg solids/kg gas and turbulent flow in the fluid stream, homogeneous suspension of the solid particles in the carrier gas is achieved relatively easily provided that the mean particle size is not very large: in practice, solids–gas loading for dilute-phase transport usually falls in the range 0.01–15 kg of solids/kg of gas.

For a required rate of solids transport, the design of a dilute-phase conveying system requires specification of the operating solids–gas mass–flow ratio, pipe diameter and the expected overall pressure drop during conveying. To avoid saltation and choking occurring during transportation, it is also important to estimate saltation and choking velocities for the system under investigation.

The prediction of the overall pressure drop and the consequential energy requirement in turn requires detailed knowledge of the fluid and particle velocities. This information is obtained from an analysis of the forces acting on the individual and cloud of particles in suspension.

Particle and fluid velocities

Consider the vertical motion of a gas–solid suspension of length dL and radius R (Figure 5.4). For fully accelerated flow in a long pipe the forces acting on the particles in suspension are drag, dF_d, friction, dF_f, and gravity, dF_g. Thus, for steady flow a force balance on the elemental slice gives:

$$dF_d - dF_f - dF_g = 0 \tag{5.1}$$

(i) Drag force

A particle falling freely at its terminal velocity in a stationary fluid will experience a resisting force due to the pressure of the fluid. This force is known as the drag force and for engineering purposes is usually expressed in terms of a single-particle drag coefficient, C_d, which is obtained experimentally. Thus:

$$F_d = C_d A \rho_b \frac{U_t^2}{2} \tag{5.2}$$

where A is the particle projected area in the direction of flow. U_t is the particle terminal velocity which, in this case, is equal to the observed particle velocity U_p since fluid velocity is zero. Equating the upward drag force to the net downward gravitional force gives:

$$U_t = \left[\frac{2w_p g(\rho_p - \rho_f)}{C_d \rho_p \rho_f A} \right]^{1/2} \quad (5.3)$$

where w_p is particle mass. For spherical particles Equation 5.3 may be rewritten as;

$$U_t = \left[\frac{4}{3} \frac{g d_p (\rho_p - \rho_f)}{\rho_f C_d} \right]^{1/2} \quad (5.4)$$

To evaluate U_t, the value of the drag coefficient is needed. C_d is, however, a function of the particle Reynolds number defined as:

$$Re_p = \frac{d_p \rho_f U_t}{\mu_f} \quad (5.5)$$

which in turn requires U_t for its evaluation.

There are several methods available that overcome this difficulty in determining U_t. One such approach which is convenient in practice is from McCabe and Smith (1976) and involves the evaluation of the following parameter:

$$K = d_p \left[\frac{g \rho_f (\rho_p - \rho_f)}{\mu_f^2} \right]^{1/3} \quad (5.6)$$

In the Stokes' flow region:

$K < 3.3$

and:

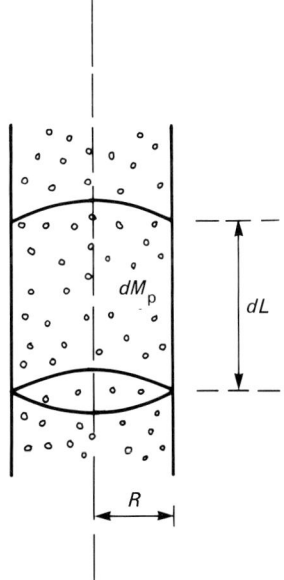

Figure 5.4 Schematic of vertical differential balance.

$$U_t = \frac{gd_p^2(\rho_p - \rho_f)}{18\mu_f} \tag{5.7a}$$

while for the intermediate regime:

$3.3 < K < 43.6$ and:

$$U_t = \frac{0.153 g^{0.71} d_p^{1.14}(\rho_p - \rho_f)^{0.71}}{\rho_f^{0.29} \mu_f^{0.43}} \tag{5.7b}$$

In Newton's region:

$K > 43.6$ and:

$$U_t = 1.75 \left[\frac{gd_p(\rho_p - \rho_f)}{\rho_f} \right]^{1/2} \tag{5.7c}$$

In a multi-particle flow system, the drag force on each particle is increased by the presence of neighbouring particles. This effect is often expressed in terms of the free space available between the particles, that is, the void fraction, ϵ. For pneumatic transport systems the suggested relationship between single- and multi-particle drag coefficients is of the form (Wen and Yu, 1966; Wen and Galli, 1971).

$$\frac{C_{dm}}{C_d} = \epsilon^n \tag{5.8}$$

where the recommended value for the exponent n is -4.7.

Thus, for spherical multi-particle flow systems, combining Equations 5.4 and 5.8 gives:

$$U_t = \left[\frac{4}{3} gd_p \frac{\Delta\rho}{\rho_f} \frac{\epsilon^{4.7}}{C_d} \right]^{1/2} \tag{5.9}$$

All the equations presented so far are only applicable if the fluid is stationary. For pneumatic conveying systems the fluid is moving, and the velocity term in Equations 5.2–5.8 should be replaced by the particle slip velocity, U_s, which is defined as the fluid velocity, U_f, minus the particle velocity, U_p, that is:

$$U_s = U_f - U_p \tag{5.10}$$

Thus, for spherical multi-particle flow systems, the drag force may be obtained by modifying Equation 5.2 using Equations 5.8 and 5.9 to give the total drag force acting on the particles in the section dL of the transport line:

$$dF_d = \frac{3}{4} C_d \epsilon^{-4.7} \left(\frac{\rho_f}{\rho_p} \right) \left[\frac{(U_f - U_p)^2}{gd_p} \right] \left\{ \frac{M_p dL}{U_p} \right\} \tag{5.11}$$

where the term in $\{\ \}$ represents the weight of the particles in the differential slice.

(ii) Frictional force

The frictional force on the particles due to particle–particle and particle–wall collisions is usually expressed in terms of a particle coefficient of friction, f_p (Yang, 1973):

$$dF_f = f_p \left[\frac{U_p^2}{2D} \right] \left\{ \frac{M_p dL}{U_p} \right\} \tag{5.12}$$

(iii) Gravitional force

For vertical pipes, this is:

$$dF_g = \left\{ \frac{M_p dL}{U_p} \right\} g \tag{5.13}$$

Substituting for dF_d, dF_f, and dF_g (Equations 5.11, 5.12 and 5.13) into Equation 5.1 and rearranging yields (Yang, 1973):

$$U_p = U_f - \left[\left(1 + \frac{f_p}{D} \times \frac{U_p^2}{2g} \right) \frac{4}{3} \times \frac{(\rho_p - \rho_f) \, d_p g \epsilon^{4.7}}{\rho_f C_d} \right]^{1/2} \tag{5.14}$$

For horizontal transport, the drag force, dF_d, and particle frictional force, dF_f, are identical to Equations 5.11 and 5.12, but the gravitational force, dF_g, is different as shown below.

A particle with a horizontal velocity U_p and terminal settling velocity, U_t, moving through a horizontal distance dL will fall vertically through a distance $U_t(dL/U_p)$ due to the gravitational force (Wen, 1976). To keep the particle in suspension, the carrier gas must provide an upward force equal to dF_g to lift the particle. dF_g is thus given by (Wen, 1976):

$$dF_g = U_t g \left\{ \frac{M_p dL}{U_p} \right\} \tag{5.15}$$

where U_t is the multi-particle terminal velocity in stagnant fluid given by Equation 5.9.

Substituting Equations 5.11, 5.12, and 5.15 into Equation 5.1 and rearranging gives the relationship between particle velocity and fluid velocity for horizontal pneumatic conveying systems. Thus:

$$\frac{3}{4} \frac{C_d \epsilon^{-4.7}}{g d_p} \frac{\rho_f}{\rho_p} (U_f - U_p)^2 = \left[\frac{4}{3} \frac{g d_p}{U_p^2} \frac{\Delta \rho}{\rho_f} \frac{1}{C_d \epsilon^{-4.7}} \right]^{1/2} + \left[\frac{f_p U_p^2}{g 2D} \right] \tag{5.16}$$

The use of Equations 5.14 and 5.16 requires knowledge of the particle friction factor, f_p. Table 5.2 provides literature values for f_p for some materials (Yang, 1974), while Table 5.3 gives a number of published correlations for estimating f_p. For detailed information on methods of measuring f_p, the original papers should be consulted.

Yang (1973) compared experimental values of Hariu and Molstad (1949) for particle velocity with those calculated using Equation 5.14 for vertical

112   Pneumatic conveying of bulk solids

pneumatic pipelines. A total of 116 data points were tested for comparison, and more than 96% of the calculated values fell within ± 20% of the observed values.

Wen (1976) tested the applicability of Equation 5.16 for horizontal flow lines using experimental data of Richardson and McLeman (1960); very good agreement was observed between experimental and calculated values of U_p.

One shortcoming of Equations 5.14 and 5.16 is that U_p is expressed implicitly. Hinkle (1953) recommended an empirical equation based on photographic measurements of particle velocity in 2 in. and 3 in. internal diameter glass pipes. This is:

$$U_p = U_f [1 - 0.68 d_p^{0.92} D^{-0.54} \rho_f^{-0.2} \rho_p^{0.5}] \tag{5.17}$$

Klinzing (1981) claims that the following Equation, from IGT (1978), gives a better fit of the experimental data:

$$U_p = U_f [1 - 0.004 d_p^{0.3} \rho_p^{0.5}] \tag{5.18}$$

Table 5.2 Particle friction factor for various materials

Material	Particle friction factor, f_p	Worker
Polystyrene	0.008–0.019	Hinkle (1953)
Coal and coke	0.005	Barth (1960)
Sea sand	0.008–0.019	Hariu and Molstad (1949)
Ground cracking catalyst	0.008–0.018	" "
Microspherical cracking catalyst	0.008–0.023	" "
Rice	0.0058	Ikemori (1959)
Soybean	0.0081	" "

Table 5.3 Equations for particle friction factor, f_p

Equation for f_p	Worker
$\dfrac{0.2}{U_p}$	Leung and Wiles (1976)
$\dfrac{0.114 \sqrt{gD}}{U_p}$	Konno and Saito (1969)
$0.0126 \left(\dfrac{1-\epsilon}{\epsilon^3} \right) \left[\dfrac{(1-\epsilon)Re}{Re_s} \right]^{-0.979}$	Yang (1978)
$0.117 \left[\dfrac{1-\epsilon}{\epsilon^3} \right] \left[\dfrac{(1-\epsilon)Re_p U_t}{\sqrt{gD}\, Re_s} \right]^{-1.15}$	Yang (1974)

for horizontal pipes and $Re_s = \rho_f d_p (U_f - U_p)/\mu_f$

$\left[0.1006 \left(\dfrac{M_p}{\rho_f U_p} \right)^{0.0415} \left(\dfrac{U_p}{U_f} \right)^{-0.859} \right] - 0.12$	Knowlton and Bachovchin (1975)

for pressures up to 200 p.s.i.a.

Equations 5.17 and 5.18 together with some other published correlations for U_p are summarized below.

Correlations for particle velocity

U_p	Ref
$U_f(1 - 0.68 d_p^{0.92} D^{0.54} \rho_p^{0.5} \rho_f^{-0.2})$	Hinkle (1953)
$U_f(1 - 0.004 d_p^{0.3} \rho_p^{0.5})$	IGT (1978)
$U_f - U_t[1 + (f_p U_p^2/2gD)]^{\epsilon^{4.7}}$	Yang (1984)
$U_f - U_t$	Terminal velocity analysis

Pressure drop in dilute-phase pneumatic conveying

For fully accelerated flow, most investigators have assumed that the overall pressure drop during conveying can be considered as the sum of two parts, static and frictional:

$$\Delta P_T = \Delta P_S + \Delta P_F \tag{5.19}$$

The pressure drop due to the static head of bulk material, ΔP_S, is zero for horizontal pipes, and for vertical flowlines it may be split up into particles and fluid contributions according to Equation 5.20:

$$\Delta P_S = \rho_p(1 - \epsilon)Lg + \epsilon \rho_f Lg \tag{5.20}$$

due to solids: (ΔP_{S_p})

due to fluid: (ΔP_{S_f})

The extra pressure drop attributable to fluid alone is substantially smaller than that due to the solid particles and can be ignored for all practical purposes.

Likewise, the frictional pressure drop term, ΔP_F, in Equation 5.19 may be divided into two parts:

$$\Delta P_F = \frac{f_p \rho_p (1 - \epsilon) U_p^2 L}{2D} + \frac{f_f \rho_f U_f^2 L}{2D} \tag{5.21}$$

due to particles: (ΔP_{F_p})

due to fluid: (ΔP_{F_f})

f_f in Equation 5.21 is the conventional Fanning friction factor which may be obtained in the usual way (see any standard textbook in chemical engineering).

Schuchart (1968) proposed an alternative equation for the estimation of the frictional pressure drop for the transport of large particles ($d_p \simeq 1$ and 2 mm diameter) in dilute-phase systems under turbulent-flow conditions:

$$\frac{\Delta P_F}{\rho_f U_f^2/2} \frac{D}{L_{eqv}} = \left\{ \frac{2.7}{k^2} C_v' \left(\frac{D}{d_p}\right) \left(\frac{U_f}{U_p}\right) \left[1 - \frac{U_p}{U_f}\right]^2 \right\} C_d \quad (5.22)$$

Where C_d is the single-particle drag coefficient and k is a form factor having a value of 0.64 for glass particles and 0.81 for plastic beads. The ratio of particle velocity to fluid velocity is given by:

$$\frac{U_p}{U_f} = \left[1 + B\left(\frac{\rho_p}{\rho_f} - 1\right)^{0.66} \left(\frac{D}{d_p}\right)^{-0.66} \left(1 + \frac{200}{Fr - Fr_t}\right)\right]^{-1} \quad (5.23)$$

where Fr is the fluid Froude number (U_f^2/gD) and Fr_t is the Froude number representing the condition when particles begin to settle at the bottom of the pipe and is given by U_t^2/gD. B is a constant varying between 0.014 and 0.5 depending upon particle properties and the material of construction of the pipe.

Hinkle (1953) provided the following expression for the total pressure drop in vertical pneumatic conveying systems:

$$\Delta P_T = \frac{U_f^2 \rho_f}{2} + \frac{M_p U_p}{A} + \frac{2 f_f \rho_f U_f^2 L}{D} + \frac{2 f_p U_p M_p L}{DA} \quad (5.24)$$

where U_p/U_f is given by Equation 5.17 and

$$f_p = \frac{3}{2}\left(\frac{\rho_f}{\rho_p}\right)\left(\frac{D}{d_p}\right) C_d \left[\frac{U_f - U_p}{U_p}\right]^2 \quad (5.25)$$

Saltation and choking velocities

For trouble-free and economic operation of pneumatic conveying systems in dilute-phase, the design must provide information on the operating gas velocity. This should be as low as possible to minimize pipe erosion, particle breakage and energy requirement in the gas mover, but high enough to ensure that all particles remain in suspension during conveying. For a horizontal pipe this velocity is defined as the 'saltation velocity', while for a vertical conveyor it is known as the 'choking velocity' (Zenz, 1964; Yang, 1975).

In the absence of any theory, several empirical correlations have been reported in the literature for the prediction of saltation and choking velocities. Large differences exist in the form of these equations, making it difficult for the designer to choose the best expression to use in practice. In order to make this task easier, some of the better tested correlations are considered here.

For pneumatic transport in horizontal pipes, Zenz (1949) plotted the minimum suspension (saltation) velocity for single particles against particle size (Figure 5.5). The curves indicate that saltation velocity decreases to a minimum with decrease in particle size and then increases again as particle size is reduced further. This is in accord with the experimental observation that a bulk material with a wide particle size distribution is the most

Pneumatic conveying of bulk solids   115

difficult to transport pneumatically. For mixed-size particles, Zenz recommended the following empirical equation for predicting saltation velocity:

$$\frac{m_p}{\rho_p} = 0.21 \, S^{1.5} \left(\frac{U_{sl} - U_{sp}}{U_{sp}} \right) \quad (5.26)$$

for m_p in kg/m² s^{-1} and $S > 0.05$.

To evaluate S, the saltation velocities should be calculated from Figure 5.5 for the smallest and the largest particles in the mixture. The slope of the line connecting the two points gives the value of S. This is then used in Equation 5.26 together with the larger value of the two saltation velocities.

Jones and Leung (1978) pointed out that Equation 5.26 has a serious drawback since it cannot be used for fine particles for which $S < 0.05$.

Barth (1958) used the dimensionless fluid Froude number at saltation defined as U_{sp}^2/gD to obtain an expression for the saltation velocity. He noted a simple power law relationship between this parameter and the particle–gas mass ratio:

$$\frac{M_p}{M_f} = K_1 Fr_s^\beta \quad (5.27)$$

where the value of the power law index β was found to be 4.0.

Equation 5.27 was later modified by Rizk (1976) to a more general form given by:

$$\frac{M_p}{M_f} = \frac{Fr_s^{(1.1 d_p + 2.5)}}{10^{(1.4 d_p + 1.96)}} \quad (5.28)$$

with d_p in mm.

Owens (1969) proposed that saltation will occur if the Froude number, defined as:

$$Fr^* = \frac{\rho_f U^2 f_f}{2 \rho_p g d_p} \quad (5.29)$$

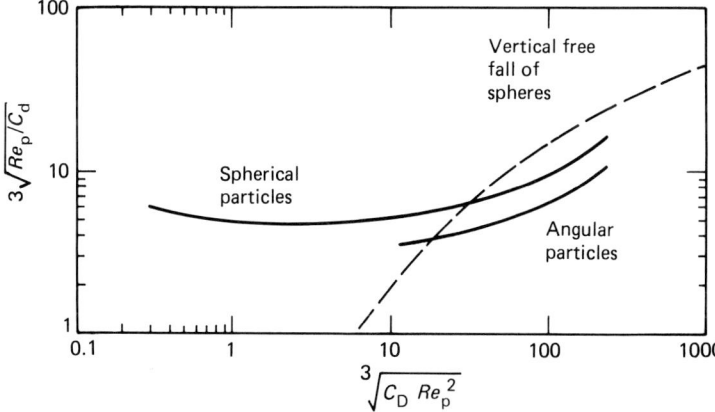

Figure 5.5 Single-particle saltation velocities (Zenz, 1964).

falls below 0.001, while for $1 < Fr^* < 0.001$ particle deposition could still be a problem.

Matsumoto et al. (1975) developed the following equation to correlate their data; it uses two Froude numbers, one for the fluid and the other for particles under settling conditions i.e. U_t^2/gD:

$$\frac{M_p}{M_f} = 0.221 \left(\frac{\rho_p}{\rho_f}\right)^{0.55} Fr_t^{-2.3} Fr_s^{3.0} \tag{5.30}$$

with an accuracy of ±40%.

Several other correlations have been reported in the literature. These were tested against published experimental data by Jones and Leung (1978). Their analysis, which is based on a total of 390 data points reveals that Rizk's Equation 5.28 and Matsumoto's correlation (Equation 5.30) are considerably more accurate for the estimation of saltation velocity than are other correlations. Observing that Equation 5.30 provides no significant improvement in the estimation of saltation velocity compared to Equation 5.28, Jones and Leung (1978) recommended that design should be based on Equation 5.28, since it is more convenient to use.

For vertical transport systems, Yang (1973) proposed that choking is avoided provided the following criterion is met:

$$\frac{U_t^2}{gD} < 0.35 \tag{5.31}$$

while for high-pressure systems, Knowlton and Bachovchin (1975) suggested the following expression for choking velocity, U_{ck}:

$$\frac{U_{ck}}{\sqrt{gD}} = 9.07 \left(\frac{\rho_p}{\rho_f}\right)^{0.347} \left(\frac{M_p d_p}{\mu_f}\right)^{0.214} \left(\frac{d_p}{D}\right)^{0.246} \tag{5.32}$$

Other reported correlations for U_{ck} are given in Table 5.4, page 126.

Chong and Leung (1986) have examined most of the available data on choking using Geldart's classification. These workers concluded that the empirical correlation reported by Yousifi and Gau (1974) is suitable for Geldart Group A and B powders while Yang's equation (Table 5.4 on page 126) is recommended for Group D materials.

Pressure drop through bends and fittings

The analysis presented so far has been simplified considerably by assuming fully accelerated flow. This assumption is reasonable for conveying in long, straight pipelines, but not for flow in short standpipes, bends, flow diverters, valves, gates, expansions and other fittings. Substantial changes occur in fluid and particle velocities whenever such items are fitted in the flowline. For example, flow through a bend is accompanied by a deceleration as the particles hit the walls of the bend. To prevent settling and consequently blockage in the flowline, the suspension must be brought back to its original state after passage through the obstruction. This is done by accelerating the particles downstream of the obstruction.

Moreover, experimental data suggest that the type and geometry of the obstruction can have a significant influence upon the degree of saltation (Patterson, 1959; Bowers and Wright, 1978). For example, for 90° bends, Patterson's data for pneumatic transport of coal particles indicate that the worst geometries are vertical and horizontal bends with downward flow from the vertical to the horizontal; vertical bends with upward flow from the vertical to the horizontal provided the best geometry with regards to saltation.

Rose and Duckworth (1969) conducted a series of studies on the pneumatic transport of particulate solids in pipes of different angles of inclination. Based on this investigation they proposed Equation 5.33 for estimating the initial acceleration length:

$$L_{acc} = 6D \left[\frac{M_p}{\rho_f \sqrt{g} D^{2.5}} \sqrt{\frac{D}{d_p} \frac{\rho_p}{\rho_f}} \right]^{1/3} \tag{5.33}$$

and the following equation for the pressure drop in that region:

$$\Delta P_{acc} = 1.12 \left(\frac{\rho_f U_f^2}{2} \right) \left(\frac{M_p}{M_f} \right) \varphi_1 \varphi_2 \tag{5.34}$$

where $\varphi_1 = fn(\rho_p/\rho_f)$

and $\varphi_2 = fn(\theta)$

Rose and Duckworth provided chart solutions for the evaluation of φ_1 and φ_2. For design calculations, these may be represented adequately by the following equations:

$$\varphi_1 = (\rho_p/\rho_f)^\alpha \tag{5.35}$$

where $\alpha = 1.0$ for $\frac{\rho_p}{\rho_f} > 10$

and $\alpha = 1.6$ for $1.8 < \frac{\rho_p}{\rho_f} < 10$

while $\varphi_2 = 1.0$ for $\theta > 60°$

and $\varphi_2 = 0.14(\theta)^{0.4}$ for $60° < \theta < 90°$

Schuchart (1968) proposed the following equation:

$$\frac{\Delta P_{bend}}{\Delta P_{F_p}} = 210 \left[\frac{D}{2R_b} \right]^{1.15} \tag{5.36}$$

where ΔP_{F_p} is the pressure drop for an equivalent length of straight pipe given by Equation 5.22 and R_b is the radius of the bend. In order to minimize ΔP_{bend}, the radius of the bend should be at least 6 to 12 times pipe diameter (Wen, 1976).

For pipe fittings, valves, gates, bends, etc., Klinzing (1981) recommended that in the absence of any fundamental work, pressure drop may be evaluated using the concept of equivalent length of straight pipe which is well established for single-phase flow.

Most of the published work to date on pneumatic conveying of particulate solids has been confined to mono-size or a narrowly classified mixture of particles. However, some experimental data have been reported (Peters and Klinzing, 1972; Pfeffer and Rossetti, 1972; Soo and Trezek, 1966) which indicate a significant reduction in pressure drop with the addition of fine particles (20–75 μm) to the mixture for solids loadings in the range 0.5–4 under fully developed turbulent flow conditions (15 000 < Re_f < 40 000). The pressure reduction that accompanies the addition of fine particles to the suspension is believed to be due to a complex interaction between fine particles and fluid turbulence, particularly close to the walls of the conveyor pipe (Klinzing, 1981). However, the phenomenon is still poorly understood and further basic research is needed in this area before its beneficial effects can be realized in industrial situations.

In contrast to this reduction in pressure drop due to the presence of fine particles, large increases in the overall pressure drops have been reported by a few investigators when conveying solid particles continuously over prolonged periods of time. This effect has been attributed to electrostatic charges between the particles (Klinzing, 1981) and between the particles and the pipe walls (Richardson and McLeman, 1966). Pressure-drop measurements in pneumatic conveying systems of electrostatically charged particles have also been reported by Joseph and Klinzing (1983) and Klinzing (1981), but the phenomenon is far from clear.

Finally, most of the systematic experimental work to date has been confined to small-scale equipment and at present it is difficult to formulate rules for scale-up purposes. For that reason design of industrial pneumatic transport systems for new materials still depends largely on trial and error using large-scale rigs.

Dense-phase pneumatic conveying systems

As gas velocity decreases below the saltation velocity, particles begin to precipitate and either move along the pipe by rolling and sliding or form into dunes with individual grains jumping from one dune to the next in the direction of fluid flow. With dense-phase flow, solids–gas loading usually falls between 15 and 200 kg of solids/kg of gas.

This method of transportation has found many applications in the past years because of the high solids–gas ratios that can be handled and the consequential savings in cost and space that can result from it. In many cases, such considerations override the main disadvantage of dense-phase flow, which is the high pressure drop compared to dilute-phase conveying.

The behaviour of solids motion in dense-phase flow is more complex in comparison with dilute-phase transport. For example, in the vertical flow direction, flow regime can change from dense-phase slugging to dense-phase without slugging and to moving bed flow with alteration in gas velocity and solids loading (Leung and Wiles, 1976). In horizontal flow, on the other hand, the formation, size and separation of dunes is affected significantly by gas velocity, pipe diameter, and solids loading (Wen and Simons, 1959). Moreover, solids velocities vary considerably locally and

Pneumatic conveying of bulk solids   119

from point to point as particles are decelerated and accelerated alternately in moving from one dune to another.

Such flow behaviour makes any theoretical analysis of dense-phase conveying extremely difficult. Consequently, reliable design information is sparse and tentative, with large safety factors used commonly, to avoid disappointment during operation.

For dense-phase conveying in horizontal pipes, Wen and Simons (1959) and Wen (1976) pointed out that the main contribution to the pressure drop is due to friction between the particles and between the particles and the gas, i.e., the frictional pressure drop due to the flow of the gas alone is negligible. Moreover, they suggest that as flow is practically independent of individual particle properties such as size and shape, it is more meaningful to consider the particle velocity as the average velocity of the mass of particles over the entire length of the pipe calculated using the dispersed solids density, ρ_s. These workers developed the following empirical correlation for estimating the pressure drop:

$$\left(\frac{\Delta P_T}{L\rho_s}\right)\left(\frac{D}{d_p}\right)^{0.25} = 2.5(U_p)^{0.45} \tag{5.37}$$

and Equation 5.38 for the evaluation of the average particle velocity:

$$U_p = \frac{m_p}{\rho_s} \tag{5.38}$$

Equation 5.37 was checked against data obtained by several other investigators (Wen, 1976) covering particle sizes in the range 70–1670 μm and for solids loading ratios between 25 and 900.

For dense-phase flow in the vertical direction, Leung and Wiles (1976) assume that the overall pressure drop is given essentially by the weight of the particles in the transport line, that is, wall frictional losses are neglected. Thus:

$$\Delta P = \rho_p(1 - \epsilon)gL \tag{5.39}$$

For dense-phase slugging flow, Leung and Wiles (1976) recommended the use of the following expression for estimating ϵ:

$$\frac{1 - \epsilon}{1 - \epsilon_0} = \left[U_b + \frac{M_p}{\rho_b(1 - \epsilon_0)}\right] \bigg/ \left(U_{sf} + U_b - U_0 + \frac{M_p}{\rho_p}\right) \tag{5.40}$$

where U_b = bubble rise velocity in a non-flowing bed and is equal to $0.35\sqrt{gD}$.

No comparable equation exists for the estimation of ϵ for dense-phase flow without slugging. Leung and Wiles (1976) suggest that in such a case, to a first approximation, a voidage of about 0.6–0.8 may be assumed, depending on the operating conditions.

For the moving-bed flow regime, Leung and Wiles suggest a modified Ergun type expression for the estimation of the pressure drop:

120 Pneumatic conveying of bulk solids

$$\frac{\Delta P d_p \epsilon^3}{L\rho_f[(U_f - U_p)]^2(1 - \epsilon)} = \frac{150}{Re_{sl}} + 1.75 \tag{5.41}$$

where $Re_{sl} = \dfrac{\rho_f(U_f - U_p)d_p\epsilon}{\mu_f(1 - \epsilon)}$

It is important to realize that at this stage of development all the equations presented for dense-phase flow are tentative and should therefore be treated with caution.

Two industrial dense-phase conveying systems for bulk solids are shown in Figure 5.6a,b and 5.7.

With the basic blow-tank arrangement, the sequence of operation starts when the heavy-duty inlet valve of the hopper opens and material falls in until a preset level is reached: the valve closes cutting through a head of material. The vessel is then pressurized by air fed into the top, and as a result the bulk material discharges down the pipeline in single-slug configuration. Although the operation of the blow tank is batchwise, by using more than one tank continuous output may be achieved quite easily (Figure 5.8).

In the pulse-phase type, the bulk material is fed and the vessel is pressurized as before. However, with pulse-phase conveying, air is fed to an aeration probe and ring positioned in the core of the hopper. This allows a dense flow of the material to pass into the pipeline. Beyond the discharge valve is located an air knife which injects pulses of air into the conveying line, and as a result the bulk material is divided into discrete plugs (Figure 5.7). When the vessel is emptied of its contents, the tank pressure is reduced to atmospheric and the cycle is repeated. A modification of pulse-phase conveying involves bypassing some of the air around the solids in order to reduce plugging of the pipeline which might occur when conveying materials over long distances.

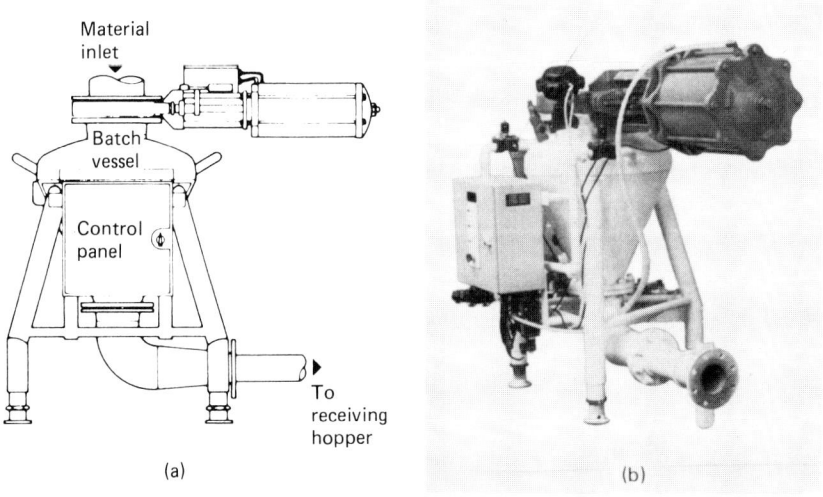

Figure 5.6(a,b) Blow bank for dense-phase conveying.

Pneumatic conveying of bulk solids 121

In practice, the choice between the simple dense-phase, the pulsed-phase and the air bypass pulsed-phase systems depends largely upon the type of material and the distance to be covered. Knowlton (1986) recommends the rather empirical procedure originally proposed by Canning and Thompson, which is based upon Geldart's fluidization classification of powders. The recommendation states that Group C

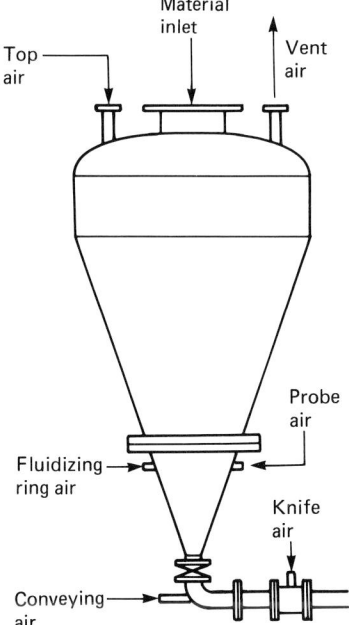

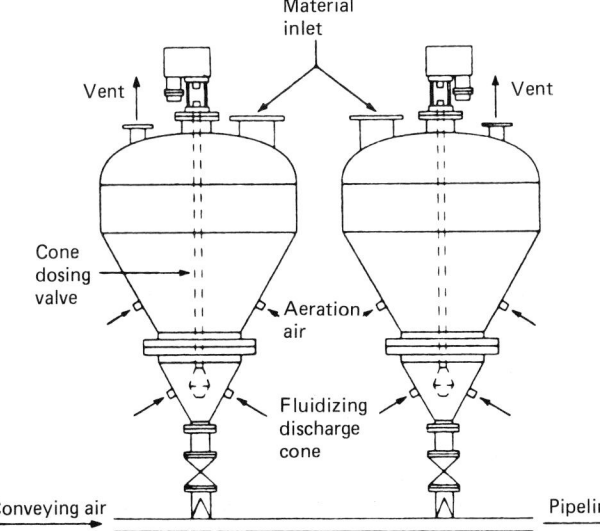

Figure 5.7 Operation of a pulse-phase conveying unit.

Figure 5.8 Multidense-phase conveying arrangement of blow tanks.

powders are generally extremely difficult to transport in dense-phase because of the cohesive nature of these materials. Group A powders are, on the other hand, very easy to convey in dense phase, and transportation may be carried out at relatively low velocities and at high solids–gas loadings. Group D powders may be transported in dense phase, provided that the material has a relatively narrow particle size distribution. Group B materials pose considerable difficulties during transportation, causing pipe vibration and blockage.

References

BARTH, W. (1960). *Chem.-Ing. Technik*, **30**, 171–180
BOWERS, H. M. and WRIGHT, J. G. (1968). Pneutransport 4, organized by BHRA Fluid Eng., California, USA, A2-7–A2-20
BRAGG, G. M. and KWAN, M. Y. M. (1978). Pneutransport 4, organized by BHRA Fluid Eng., California, USA, C2-13–C2-26
CHONG, Y. O. and LEUNG, L. S. (1986). *Powder Technol.*, **47**, 43–50
DALLA VALLE, J. M., (1942 Nov.). *Heating and Ventilating*, **39**, 28–32
DOIG, I. D. and ROPER, G. H. (1963). *Aust. Chem. Eng.*, **4**, 9
HINKLE, B. L. (1953). PhD thesis, Georgia Inst. of Technol., Atlanta, Georgia, USA
HARIU, O. H. and MOLSTAD, M. C. (1949). *Ind. Eng. Chem.*, **41**, 1148–1160
HUGGETT, M. R. and GODFERY, A. R. (1977). *Powtech 77, UK*, 18–27
IGT (Institute of Gas Technology), (1978). Dept. of Energy Contact, FE 2286-32, USA
IKEMORI, K. (1959). *J. Mech. Eng. Japan*, **62**, 480
JONES, P. S. and LEUNG, L. S. (1978). Pneutransport 4, organized by BHRA Fluid Eng., California, USA, C1-1–C1-12
JOSEPH, S. and KLINZING, G. E. (1983). *Powder Technol.*, **36**, 79–87
KLINZING, G. E. (1981). *Gas–SolidsTransport*, McGraw-Hill, New York
KNOWLTON, T. M. (1986) in *Gas Fluidization Technology*, Geldart, D. (Ed.). Interscience (John Wiley). pp. 341–415
KNOWLTON, T. M. and BACHOVCHIN, D. M. (1975) (June). Engng. Foundation Conf. on Fluidization, Asilomar
KONNO, H. and SAITO, S. (1969). *J. Chem. Eng. Japan*, **2**, 211–217
KRAUS, M. N. (1980). *Pneumatic Conveying of Bulk Solids*, 2nd ed. McGraw-Hill
LEUNG, W. C. and WILES, R. J. (1976). *Ind. Eng. Chem. Proc. Des.*, 552–557
LEUNG, L. S., WILES, R. J. and NICKLIN, D. J. (1971). *Ind. Eng. Chem.: Proc. Des. Dev.*, **10**, 183–189
MCCABE, W. L. and SMITH, J. C. 9176). *Unit Operation of Chemical Engineering*, 3rd End. McGraw-Hill, New York
MASON, J. S. (1977). *Powtech 77, UK*, 1–5
MATSEN, J. M. (1982). *Powder Technol.*, **32**, 21–33
MATSUMOTO, S., HADDA, S., SAITO, S. and MAEDA, S. (1975). *J. Chem. Eng. Japan*, **7**, 331–333
OWENS, P. R. (1969). *J. Fluid. Mech.*, **39**, 407–432
PATTERSON, R. C. (1959). *Trans. Am. Soc. Mech. Engrs.*, **81**, No. 43, 43–54
PERRY, R. H. and CHILTON, C. H. *Chemical Engineering Handbook*, 5th Edn, McGraw-Hill
PETERS, L. K. and KLINZING, G. E. (1972). *Can J. Chem. Eng.*, **50**, 441–444
PFEFFER, R. and ROSSETTI, S. J. (1972). *AIChE*, **18**, 31–39.
SOO, S. L. and TRESEK, G. J. (1966). *Ind. Eng. Chem; Fund.*, **5**, 388–392
RICHARDSON, J. F. and MCLEMAN, M. (1960). *Trans. I. Chem. E.*, **38**, 257–266
RIZK, F. (1976). Meeting organized by BHRA Fluid Eng., Cranfield, UK., paper D4
ROSE, H. E. and DUCKWORTH, R. A. (1969) (March). *The Engineer*, 478
SCHUCHART, P. (1968). *Chem.-Ing. Technol.*, **40** (21/22), 1968

STEPHANOFF, A. J. (1969). *Gravity Flow of Bulk Solids and Transportation of Solids in Suspension*, John Wiley
STOESS, H. A. JR. (1983). *Pneumatic Conveying*, John Wiley
WEN, C. Y. (1976). In *Gas-solids Handling in the Process Industries*, Marchello, J. M. and Gomezplata, A. (Eds.), Marcel Dekker Inc., 89–134
WEN, C. Y. and GALLI, A. F. (1971). In *Fluidization*, Davidson, J. F. and Harrison, D. (Eds.). Academic Press, New York, 677–710
WEN, C. Y. and SIMONS, H. P. (1959). *AIChE*, **5**, 263–267
WEN, C. Y. and YU, Y. H. (1966). *Chem. Eng. Prog. Symp. Series*, No. 62, 101
WHETTON, J. T. and BROADHURST, P. M. (1952) (Aug). *Trans. Int. Mining Engrs*, **3**, 120–126
YANG, W. C. (1973). *Ind. Eng. Chem; Fund.*, **12**, 349–352
YANG, W. C. (1974). *AIChE*, **20**, 605–607
YANG, W. C. (1975). *AIChE*, **21**, 1013–1015
YANG, W. C. (1978). *AIChE*, **24**, 548–552
YANG, W. C. (1983). *Powder Technol.*, **35**, 143–150
YANG, W. C. (1984). *AIChE*, **30**, 1025–1027
YOUSFI, Y. and GAU, G. (1974). *Chem. Eng. Sci.*, **29**, 1939–1953
ZENZ, F. A. (1949). *Ind. Eng. Chem.*, **41**, 2801–2806
ZENZ, F. A. (1964). *Ind. Eng. Chem.; Fund.*, **3**, 65–75

Symbols

A	particle projected area and pipe cross-sectional area
B	constant in Equation 5.23
C_d	single-particle drag coefficient
C_{dm}	multi-particle drag coefficient
C_V	$= \dfrac{\text{solid volumetric flow rate}}{\text{total volumetric flow rate}}$
D	pipe diameter
d_p	particle diameter
dF_d	drag force on particles in the elemental slice
dF_f	frictional force of particles in the elemental section
dF_g	gravity force acting on particles in the elemental slice
dL	length of the elemental section
f_f	fluid friction factor
f_p	particle friction factor
Fr	fluid Froude number (U_f^2/gD)
Fr_t	fluid Froude number at settling point (U_{t2}^2/gD)
Fr_s	fluid Froude number at saltation point (U_{sp}^2/gD)
Fr^*	modified fluid Froude number defined by Equation 5.29
g	acceleration due to gravity
K	constant in Equation 5.27
K	parameter defined by Equation 5.6
k	constant in Equation 5.22
L_{eqv}	equivalent length of pipe
L_{acc}	acceleration length
L	pipe length
M_p	particles mass-flow rate
M_f	fluid mass-flow rate

m_p particles mass-flow rate per unit cross-sectional area
m_f fluid mass-flow rate per unit cross-sectional area
ΔP_T total pressure drop during transport
ΔP_S pressure drop due to the static head of fluid and particles
ΔP_F frictional pressure drop
ΔP_{S_p} pressure drop due to the static head of particles
ΔP_{S_f} pressure drop due to the static head of fluid
ΔP_{F_p} frictional pressure drop due to the particles
ΔP_{F_f} frictional pressure drop due to the fluid flow alone
ΔP_{acc} pressure drop in the acceleration length
R pipe radius
R_b radius of pipe bend
Re_p particle Reynolds number ($\rho_f\, U_t\, dp/\mu_f$)
Re_f fluid Reynolds number ($\rho_f\, U_f\, D/\mu_f$)
Re_{sl} Reynolds number defined by Equation 5.41
U_b bubble velocity
U_0 superficial minimum fluidization velocity
U_t particle free-fall velocity
U_p particle velocity
U_f fluid velocity
U_{sf} superficial fluid velocity
U_{sp} single-particle saltation velocity
U_{sl} multi-particle saltation velocity
U_{ck} choking velocity
U_s particle slip velocity
w_p mass of single particle

β constant in Equation 5.27
ϵ voidage
ϵ_0 voidage at minimum fluidization velocity
θ angle of inclination (degrees)
μ fluid viscosity
ρ_p particle density
ρ_f fluid density
ρ_s suspension density
$\left.\begin{array}{l}\varphi_1\\ \varphi_2\end{array}\right\}$ functions in Equation 5.34

Example

Coal powder with a mean size of 100 μm is to be transported pneumatically from a storage silo to a fluidized-bed reactor. The pipeline arrangement is shown schematically in Figure 5.9. Determine the energy loss for this transportation given the following information:

$\mu_f = 1.0\ 10\ \text{kg/m s}$
$\rho_f = 1.2\ \text{kg/m}^3$
$\rho_p = 1300\ \text{kg/m}^3$
required solid flow rate = 0.02 kg/s
gas velocity, U_f = 7 m/s
radius of bends (CD and EF) = 0.8 m

Pneumatic conveying of bulk solids   125

Use the terminal velocity analysis to evaluate particle velocity and Leung–Wiles equation for the particle friction factor. The gas friction factor, f_f, may be assumed constant with a value of 0.07.

Solution

To calculate the overall pressure drop, it is necessary to establish whether the acceleration length and hence Δp_{accel} is important. This is done by calculating L_{accel} using Equation 5.33:

$$L_{accel} = 6D \left[\frac{M_p}{\rho_f \sqrt{g} D^{2.5}} \frac{D}{d_p} \frac{\rho_p}{\rho_f} \right]^{1/3}$$

$$= 6 \times 0.0254 \left[\frac{0.02}{1.2 \times 9.81 \times (0.025)} \left(\frac{0.0254}{100 \times 10^{-6}} \right) \left(\frac{1300}{1.2} \right) \right]$$

$$= 4.6 \text{ m}$$

Thus, for the first 4.6 m of the pipe, section AB, the pressure drop is almost entirely due to the acceleration of the particles: this is calculated using Equation 5.34:

$$\Delta p_{accel} = 1.12 \frac{\rho_f U_f^2}{2} \frac{M_p}{M_f} \varphi_1 \varphi_2$$

$$= 1.12 \left(\frac{7^2 \times 1.2}{2} \right) \times 4.6 \times 1 \times 1$$

$$= 138 \text{ N/m}^2/\text{m}$$

Thus for $L_{accel} = 4.6$ m

$\Delta p_{accel} = 634$ N/m

Assuming no loss in velocity for flow through the bends, the pressure drop for flow through the straight sections BC and DE are calculated using Equations 5.19, 5.20 and 5.21:

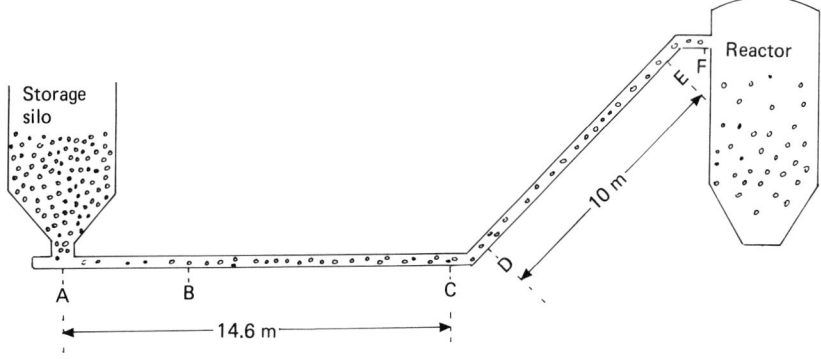

Figure 5.9 Details of the pneumatic transport.

$$\Delta p_{\text{straight}} = \rho_p(1-\epsilon)Lg + \rho_f\epsilon Lg + \frac{f_p\rho_p(1-\epsilon)U_p^2 L}{2D} + \frac{\rho_f f_f \epsilon U_f^2 L}{2D}$$

To use this equation, U_p and f_p must first be calculated. Particle velocity is given by:

$$U_p = U_f - U_t$$

where U_t is the particle free-fall velocity which may be obtained using the K criterion. Thus

$$K = d_p\left(\frac{g\rho_f(\rho_p - \rho_f)}{\mu_f^2}\right)^{1/3}$$

$$= 100 \times 10^{-6}\left[\frac{9.81 \times 1.2 \times (1300 - 1.2)}{(1 \times 10^{-5})^2}\right]^{1/3}$$

$$= 5.28$$

which is in the intermediate regime.

Table 5.4 Correlations for the choking velocity

Correlation	Reference
$\dfrac{U_{ck}}{U_t} = 32.3\, U_p + 0.97$	Leung et al. (1971)
$\dfrac{U_{ck}}{gd_p} = 32\, Re_p^{-0.06}\left(\dfrac{M_p}{M_f}\right)^{0.28}$ (Re_p given by Equation 5.5)	Yousfi and Gau (1974)
$\dfrac{U_{ck}}{U_t} = 1 + \dfrac{U_p/U_t}{1+\epsilon}$ $\dfrac{2gD(\epsilon^{-4.7}-1)}{(U_{ck}-U_t)^2} = 6.81 \times 10^5\left(\dfrac{\rho_f}{\rho_p}\right)^{2.2}$ (ϵ at choking condition)	Yang (1983a,b)
$\dfrac{U_{ck}}{U_t} = 10.74\, U_p^{0.227}$ (for $\dfrac{U_{ck}}{U_t} > 1.29$)	Matsen (1982)
$\log\dfrac{U_t}{\sqrt{gd_p}} = 0.03 U_t + 0.25\log\left(\dfrac{M_p}{M_f}\right)$ (for $U_t < 3.05$ m/s) $\log\dfrac{U_t}{\sqrt{gd_p}} = \dfrac{U_t - 2}{28} + 0.25\log\left(\dfrac{M_p}{M_f}\right)$ (for $U_t > 3.05$ m/s)	Doig and Roper (1963)

Therefore:

$$U_t = \frac{0.153 \, g^{0.71} \, d_p^{1.14} \, (\rho_p - \rho_f)^{0.71}}{\rho_f^{0.29} \, \mu_f^{0.43}}$$

$$= \frac{0.153 \times (9.81)^{0.71} \times (100 \times 10^{-6})^{1.14} \times (1300 - 1.2)^{0.71}}{(1.2)^{0.29} \times (1 \times 10^{-5})^{0.43}}$$

$$= 0.446 \text{ m/s}$$

And thus:

$$U_p = 7 - 0.446$$
$$= 6.53 \text{ m/s}$$

Particle friction factor, f_p, is obtained using (see Table 5.4):

$$f_p = \frac{0.2}{U_p} = 0.031$$

Furthermore, the voidage, ϵ, is obtained using the following equation:

$$\epsilon = 1 - \frac{4M_p}{(\rho_p - \rho_f)\pi D^2 U_p}$$

$$1 - \frac{4 \times 0.02}{(1300 - 1.2) \times \pi \times (0.254)^2 \times 6.53}$$

$$= 0.995$$

The total pressure drop, $\Delta p_{straight}$ for the straight sections BC and DE are now calculated:

$$\Delta p_{straight} = (1300 \times 0.005 \times 10 \times 9.81) + (1.2 \times 0.995 \times 10 \times 9.81)$$

$$+ \left(\frac{0.031 \times 1300 \times 0.005 \times 6.53 \times 10}{2 \times 0.0254} \right) +$$

$$\left(\frac{0.07 \times 1.2 \times 5.53 \times 10}{2 \times 0.0254} \right)$$

$$= 3616 \text{ N/m}^2 \, (361.6 \text{ N/m}^2/\text{m})$$

Since the two sections have the same length the pressure drop will be the same. Therefore, total pressure drop for section BC and DE will be:

$$\Delta p_{straight} = 2 \times 3616 \text{ N/m}^2$$
$$= 7232 \quad \text{N/m}^2$$

It should perhaps be pointed out that particle velocity will fall in moving through the bend and, to avoid deposition, particles should be accelerated back to their original state downstream from the bend.

The pressure drop for flow of the suspension through bends, sections CD and EF, may be calculated using Equation 5.36:

$$\frac{\Delta p_{bend}}{\Delta p_{straight}} = 210 \left(\frac{D}{2R_b}\right)^{1.15}$$

Thus

$$\Delta p_{bend} = \Delta p_{straight} \times 210 \times \left(\frac{0.0254}{0.8}\right)^{1.15}$$

$$= 4\Delta p_{straight}$$

$\Delta p_{straight}$ is the pressure drop of an equivalent length of straight pipe obtained from Equation 5.22:

$$\frac{\Delta P_F}{\rho_f U_f^2/2} \frac{D}{L_{eqv}} = \left\{\frac{2.7}{k^2} C_V \left(\frac{D}{d_p}\right)\left(\frac{U_f}{U_p}\right)\left[1 - \frac{U_p}{U_f}\right]^2\right\} C_d$$

where

$$\frac{U_p}{U_f} = \left[1 + B\left(\frac{\rho_p}{\rho_f} - 1\right)^{0.66}\left(\frac{D}{d_p}\right)^{-0.66}\left(1 + \frac{200}{Fr - Fr_t}\right)\right]^{-1}$$

Thus calculating the parameters in these equations first:

$$Fr = \frac{U_f^2}{gD} = 197$$

$$Fr_t = \frac{U_t^2}{gD} = 0.8$$

$(D/d_p) = 254$
$Re_p = 5$
$C_d = 6$
$C_V = 0.0042$
$k = 0.8$
$B = 0.02$

thus

$$\frac{U_p}{U_f} = 0.91$$

and therefore

$$\frac{\Delta p_{straight}}{L_{eqv}} = 327 \text{ N/m}^2$$

which is reasonably close to the value obtained previously using Equations 5.19, 5.20 and 5.21.

Pneumatic conveying of bulk solids 129

The equivalent length of straight pipe, L_{eqv}, is given by the following equation:

$$L_{eqv} = 0.5 \times \pi \times 0.8$$
$$= 1.27 \text{ m}$$

Therefore:

$$\Delta p_{straight} = 1.27 \times 327 \text{ N/m}^2$$

and

$$\Delta p_{bend} = 4 \times 1.27 \times 237 \text{ N/m}^2$$
$$= 1661 \text{ N/m}^2$$

and since there are two bends in the pipe network, the total pressure drop for the bends is:

$$\Delta p_{bend} = 3322 \text{ N/m}^2$$

Thus overall pressure drop for flow is the summation of all the individual components.

Thus

$$\Delta p_{overall} = \Delta p_{bends} + \Delta p_{straight} + \Delta p_{accel}$$
$$\Delta p_{overall} = 3322 + 7232 + 634$$
$$= 11188 \text{ N/m}^2$$

Chapter 6
Hydraulic transport of particulate solids

Introduction

Hydraulic conveying of solid materials through pipes has progressed enormously since its beginning well over a century ago. In the civil and mining industries, millions of tons of coal, cement, sand and mineral ores are transported every year over hundreds of kilometres by suspending the particles in water and pumping the resultant mixture (Orr, 1966; Faddick, 1982). In the chemical and allied industries concerned with the processing of particulate solids, it is difficult to find any plant that does not at some stage involve the conveying of a solid material dispersed in a liquid medium; typical examples include catalyst, nuclear and polymer particles, china clay, chalk, pigments, paints and many foodstuffs.

In general, hydraulic transport of particulate solids involves two types of material, settling suspensions and non-settling slurries.

A settling suspension is usually a mixture of relatively large ($d_p > 40$ μm) and/or heavy particles in a low-viscosity liquid such as water. In the absence of sufficient fluid turbulence, solids will rapidly settle to the bottom of the conveying pipe where, depending upon the operating conditions, the particles can remain as a permanent deposit or move along the floor of the pipe as a sliding bed.

A non-settling slurry, on the other hand, is a mixture of relatively fine particles ($d_p < 30$ μm) and a liquid. In many cases solids concentrations well over 50% by weight are quite common, with the resulting mixtures often exhibiting pronounced non-Newtonian behaviour. Furthermore, since in such systems the particles show little tendency to settle, the suspension can be readily transported in laminar as well as in turbulent flow.

Settling suspensions and non-settling slurries have widely different pressure-drop–velocity characteristics (Figure 6.1) and this can sometimes be used to distinguish between the two types of flow. However, in most practical situations the above simple classification is not always satisfactory because the type of suspension is also strongly influenced by numerous other parameters such as particle size distribution and shape, the chemical nature of the medium and the presence of any additives (Ayazi Shamlou, 1984a,b).

Regardless of the type of conveying, most solid hydraulic transport plants can be divided into the following four main stages:

(i) Particle preparation

This could, and often does, involve adjusting particle size and size distribution in order to meet process requirements and to minimize transport costs. Standard methods such as crushing, grinding, melting and pelleting may be used for this purpose, depending upon the type of material handled.

(ii) Suspension preparation

This is usually carried out by mixing the particles and the conveying liquid in one or more agitated vessels.

(iii) Suspension conveying

The most common method is to pump the suspension through a circular pipeline, using one or more pumps in series depending upon process needs. Both centrifugal and positive-displacement pumps can be used for this duty. With some fine abrasive powders, the particles can be introduced into the liquid stream after its passage through the pump, thus avoiding direct contact between the particles and the fluid mover.

(iv) Solids–liquid separation

With most high-concentration, non-settling slurries, the mixture conveyed is often in its final form and can be used in the next stage of the process

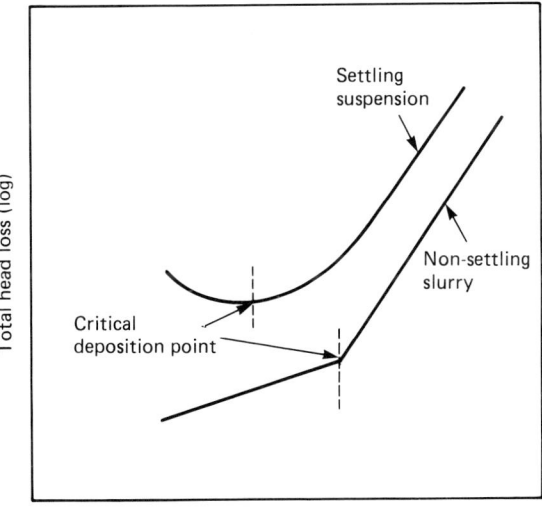

Figure 6.1 Characteristic head loss/velocity plots for settling suspensions and non-settling slurries.

without any further alteration to its form or quality. A good example of this type of suspension is super-fine coal–water slurries used as fuels for direct combustion in power station burners and boilers (Brookes and Dodwell, 1984). Settling suspensions, on the other hand, normally require further treatment after transportation. This usually involves particle classification and partial or complete dewatering, depending upon process specification.

The cost of transportation can be high, particularly when large particles are conveyed over long distances. In comparison, preparation and dewatering costs are usually low. Particles and suspension preparation, classification, and dewatering stages have been the subject of many reviews in the past and therefore will not be considered further here. For further information on these topics, the reader is referred to other relevant sources of reference (Harnby et al., 1985; Svarovsky, 1981).

Settling suspensions

From a design point of view, it is important to establish the prevailing flow pattern during transportation and to evaluate the design velocity and the associated pressure drop for flow. The conveying velocity should be high enough to prevent particles settling to the bottom of the pipe, but low enough to minimize particle breakage, pipe erosion and the specific energy requirements, i.e., power input per unit weight of particles conveyed per unit length of pipe (Streat, 1982; Lazarus, 1982).

Flow regimes

With large particles suspended in a low-viscosity liquid stream, *homogeneous* suspension is only possible if the liquid velocity is high and the flow condition is fully turbulent (Figure 6.2). With excessively large and/or heavy particles it may not be possible to achieve homogeneity regardless of fluid velocity. In such a case, a solid concentration gradient will develop across the cross-section of the carrier liquid in the pipeline and the suspension is referred to as *heterogeneous* (Figure 6.2). Most industrial hydraulic transport systems are designed to operate in this regime.

As fluid velocity is reduced, the larger grains begin to settle to the bottom of the pipe forming a layer of particles. Provided that the fluid velocity is above a critical value, the particles in this layer will move along the floor of the pipe as a *sliding bed*; in the upper section of the pipe, the particles are conveyed in the heterogeneous regime.

With further reduction in liquid velocity, more and more particles will settle towards the bottom of the pipe until eventually a permanent bed of particles is formed. This is known as saltation, and the corresponding liquid velocity is referred to as the saltation velocity.

A direct consequence of the particle deposition and saltation is a reduction in the cross-sectional area available for flow. This in turn causes an increase in the velocity of the carrier liquid. Depending on the increase in the velocity, the particles on the top of the stationary layer will be either picked up and transported back into the bulk region of the pipe or,

Hydraulic transport of particulate solids 133

alternatively, they will slide along the surface of the bed; this mode of transportation is generally referred to as sliding bed with saltation (Figure 6.2).

Pressure drop in hydraulic transport of settling suspensions

An experimental pressure-drop–velocity plot for horizontal hydraulic transport of a typical settling suspension is shown schematically in Figure 6.2. Line 1 represents frictional loss for the carrier liquid alone flowing through the pipe.

Keeping liquid velocity constant, the pressure drop increases (lines 2–4) as solid particles are introduced into the fluid stream. It is interesting to note that the particles' contribution to the overall pressure drop decreases with increase in liquid velocity until eventually the pressure drop is practically that of the fluid alone.

In deriving the relevant equations for the estimation of pressure drop for hydraulic flow of settling suspensions, most investigators have assumed, with little theoretical justification, that the total head loss may be assumed to be made up of two additive parts, i.e., $i_T = i_p + i_f$; i_f being the head loss per unit length of pipe due to flow of the fluid alone, and i_p the head loss per unit length of pipe due to the presence of solids. Newitt *et al.* (1955, 1961) suggest that such an assumption is only adequate as a first approximation since the presence of particles must interfere with the flow pattern.

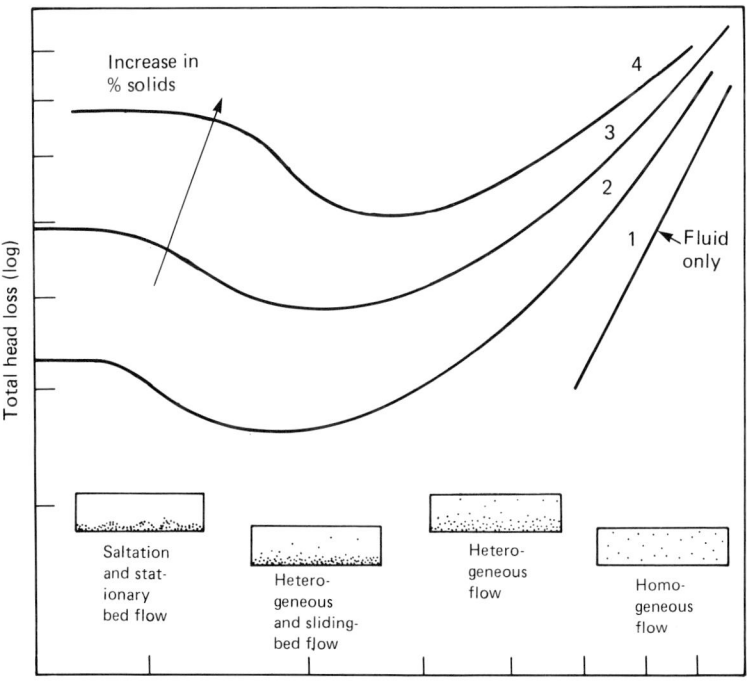

Figure 6.2 Pressure drop/velocity curves for hydraulic transport of settling suspensions.

134 Hydraulic transport of particulate solids

Durand (1953) and his co-workers (Condolios and Chapus, 1963a,b) conducted extensive investigations on the flow of solid–water suspensions in pipes. The results were treated purely empirically to provide the following correlation for the estimation of the overall pressure drop for transport in the heterogeneous and non-deposit flow regime:

$$\frac{i_T - i_f}{i_f C_v} = A \left[\frac{gD(s-1)}{V_m^2} \frac{1}{\sqrt{C_d}} \right]^m \quad (6.1)$$

with reported experimental values of $m = 1.5$. The value of the constant A in Equation 6.1 was not given in the original publication (Durand, 1953). Babcock (1971) recommended a value of 81 based on Durand's data. Stepanoff (1969) reassessed the work of Durand (Condolios and Chapus, 1963a,b) and suggested that a better fit of the experimental data may be obtained by using the numerical value of 85 for the constant A. Stepanoff's value for A was later confirmed by Turian and Yuan (1977).

The pipe frictional loss for the flow of the liquid alone, i_f, may be obtained using Darcy's relationship which is:

$$i_f = 2 f_f \frac{V_m^2}{2gD} \quad (6.2)$$

where the Fanning friction factor, f_f, may be obtained from standard charts found in most textbooks on fluid mechanics.

Durand and Condolios (1952) also investigated the critical deposit velocity for the transport of sand–water suspensions with solids concentrations, C_v, of up to 15% flowing in pipes ranging from 3.81 to 50.8 cm in diameter. They defined the critical deposit velocity as that velocity below which a permanent deposit of particles exists on the floor of the pipe and, based on their experimental results, they provided the following empirical equation for its estimation:

$$V_c = F_L 2gD(s-1) \quad (6.3)$$

where F_L is a function of particle diameter and concentration and may be obtained either graphically (Figure 6.3) or numerically using the following equation:

$$F_L = 2.43(C_v)^{1/3}/(C_d)^{1/4} \quad (6.4)$$

Equation 6.1 indicates that the contribution to the overall pressure drop due to the particles, i.e., $(i_T - i_f)$, is directly proportional to solids concentration. Moreover, the effect of particle and fluid properties such as particle size, shape and density, liquid viscosity and density are accounted for by the inclusion of the single-particle terminal settling velocity, U_t (see Chapter 5 for methods of estimating U_t). Co-workers of Durand, Condolios and Chapus (1963a,b) claim that particle shape has little effect on pressure drop and suggest that the effect of particle size distribution may be accounted for by using the mean weighted drag coefficient defined as:

$$C_{d_{wm}} = \left[\sum_{i=1}^{i=n} C_{wi} C_d \right]^2 \quad (6.5)$$

where n is the number of equal size increments and C_{wi} is the weight fraction in the increment i.

Babcock's observations (1971), however, suggest that the overall pressure drop is not directly proportional to solids concentration, and that Equation 6.1 does not describe adequately the effect of other important variables such as pipe diameter and particle properties. Subsequent investigations have also revealed that the constants A and m in Equation 6.1 should be modified for different solid–liquid suspensions (Hayden and Stelson, 1971).

For particles with specific gravity in the range 1.18–4.60, mean particle size in the range 2–600 μm and solid concentrations, C_v, of up to 37%, flowing in a 2.54 cm diameter pipe, Newitt et al. (1955) developed the following empirical equations for the various flow regimes:

$$\frac{i_T - i_f}{i_f C_v} = 0.6(s - 1) \tag{6.6}$$

for homogeneous flow,

$$\frac{i_T - i_f}{i_f C_v} = 1100(s - 1) \left(\frac{gD}{V_m^2} \right) \left(\frac{U_t}{V_m} \right) \tag{6.7}$$

for heterogeneous flow and

$$\frac{i_T - i_f}{i_f C_v} = 66(s - 1) \left(\frac{gD}{V_m^2} \right) \tag{6.8}$$

for flow in the sliding-bed regime.

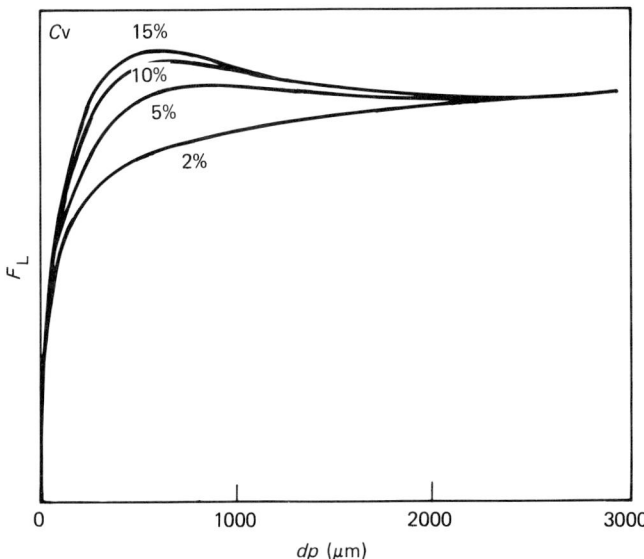

Figure 6.3 Plot of parameter F_L vs particle diameter for various solids concentrations.

136 Hydraulic transport of particulate solids

The transition velocity between homegeneous and heterogeneous flow is obtained by equating Equations 6.6 and 6.7. This yields:

$$V_{ho \to ht} = [1800gDU_t] \tag{6.9}$$

Similarly the transition velocity between heterogeneous flow and sliding-bed or saltation is obtained by equating Equations 6.7 and 6.8. This gives:

$$V_{ht \to sl} = 17U_t \tag{6.10}$$

It is interesting to note that, while $V_{ho \to ht}$ depends upon pipe diameter, $V_{ht \to sl}$ is independent of it.

Streat (1982) has reported excellent agreement between Equation 6.7 and experimental observations reported by the Saskatchewan Research Council (Haas *et al*, 1973) for the flow of sand–water suspensions in pipes ranging from 3.08 to 30.48 cm in diameter with solids concentration, C_v, of up to 25%.

Chhabra and Richardson (1983) reassessed Newitt's Equation 6.8 and claimed that a value of 59 instead of 66 would give a better fit of the experimental data.

For the vertical transport of solid–water suspensions with solids having specific gravity in the range 1.2–4.6 and mean particle diameter ranging from 100 μm to 3800 μm flowing in cylindrical pipes of 2.54 cm and 5.32 cm diameter, Newitt *et al.* (1961) also developed the following empirical expression for the estimation of pressure drop during conveying:

$$\frac{i_T - i_f}{C_v i_f} = 0.0037 \left(\frac{gD}{V_m^2}\right)^{1/2} \left(\frac{D}{d_p}\right) \tag{6.11}$$

Zandi and Govatos (1967) combined their own experimental results with data extracted from other publications in order to provide a criterion for the transition between saltation and heterogeneous flow. They assessed over 2500 experimental measurements and developed the following parameter:

$$N = \frac{V_m^2 \sqrt{C_d}}{C_v D g (s - 1)} \tag{6.12}$$

for $N > 40$, the flow is heterogeneous and pressure drop may be obtained using the following recommended correlations:

$$\frac{i_T - i_f}{C_v i_f} = 280 \, \psi^{-1.93} \tag{6.13}$$

for $\psi < 10$ and

$$\frac{i_T - i_f}{C_v i_f} = 6.3 \, \psi^{-0.354} \tag{6.14}$$

for $\psi > 10$
where

$$\psi = \frac{V_m^2}{gD(s - 1)} \sqrt{C_d} \tag{6.15}$$

At $N = 40$, saltation commences. This may be used to obtain an expression for the critical deposit velocity, V_c. Thus:

$$V_c^2 = 40 \frac{C_v g D(s-1)}{\sqrt{C_d}} \qquad (6.16)$$

Equation 6.16 and Durand's Equation 6.3 have been shown to describe test data with a fair degree of accuracy (Hayden and Stelson, 1971; White and Seal, 1982).

For $N < 40$, flow is unstable and the pressure drop is unpredictable.

Turian and Yuan (1977) also combined their own data with measurements from previously published sources, including those of Zandi and Govatos (1967) in order to develop general equations for the prediction of the pressure drop in various flow regimes. As a result of this investigation, Turian and Yuan developed the following expressions for the estimation of pressure drop for flow:

$$\frac{f_T - f_f}{f_f^{0.7717} C_v^{0.7389}} = 0.4036 \left[\frac{gD(s-1)}{V_m^2} \right]^{1.098} (C_d)^{-0.4054} \qquad (6.17)$$

for the stationary bed flow (denoted as regime 0),

$$\frac{f_T - f_f}{f_f^{1.046} C_v^{1.018}} = 0.9857 \left[\frac{gD(s-1)}{V_m^2} \right]^{1.354} (C_d)^{-0.4213} \qquad (6.18)$$

for saltation flow (regime 1),

$$\frac{f_T - f_f}{f_f^{1.2} C_v^{0.8687}} = 0.5513 \left[\frac{gD(s-1)}{V_m^2} \right]^{0.6938} (C_d)^{-0.1677} \qquad (6.19)$$

for heterogeneous flow (regime 2) and

$$\frac{f_T - f_f}{f_f^{1.428} C_v^{0.5024}} = 0.8444 \left[\frac{gD(s-1)}{V_m^2} \right]^{0.3531} (C_d)^{0.1516} \qquad (6.20)$$

for homogeneous flow (regime 3).

By equating the above equations for the various regimes, the corresponding equation for the transition velocity may be obtained. Thus, the transition velocity between regimes 1 and 3 may be obtained by equating Equations 6.18 and 6.20. This gives:

$$\frac{V_{sa \to ho}^2}{1.167(C_v)^{0.5153}(f_f)^{-0.3820}(C_d)^{-0.5724} gD(s-1)} = 1 \qquad (6.21)$$

Similarly the transition velocity between regimes 0 and 2 is:

$$\frac{V_{st \to ht}^2}{0.4608(C_v)^{-0.3225}(f_f)^{-1.065}(C_d)^{-0.5906} gD(s-1)} = 1 \qquad (6.22)$$

and the transition velocity between regimes 0 and 3 is:

$$\frac{V_{st \to ho}^2}{0.3703(C_v)^{0.3183}(f_f)^{-0.8837}(C_d)^{-0.7496} gD(s-1)} = 1 \qquad (6.23)$$

Many other correlations have been reported in the literature for the estimation of the design velocity and the prevailing pressure drop for the hydraulic transport of suspensions. In the absence of an exact theory, the majority of these equations have been developed purely empirically. However, a review of the relevant literature reveals large differences in the form of these equations, indicating that the effect of the basic variables or suspension mechanisms have not been clearly understood.

Moreover, most correlations to date have been derived for mono-size or narrowly classified particles; the effect of particle size distribution and particle shape has rarely been investigated. Classification of flow into the various flow regimes is often quite arbitrary, thus making any comparison between the various publications rather difficult.

It is therefore hardly surprising that large discrepancies exist between the predicted and the experimental values of the design velocity and pressure drop (Carleton and Cheng, 1977). Because of such uncertainties associated with most design equations, the safest way to design any hydraulic transport system is to carry out experiments on the actual material to be conveyed. Further, in order to minimize uncertainties associated with scale-up, such experimental trials should be conducted using equipment on a scale as close to the full size as is practicable (Carleton and Cheng, 1977).

Non-settling slurries

Mixtures of fine particles ($d_p < 30$ μm) in sufficiently high concentrations often exhibit little tendency to particle deposition because of strong particle–particle and particle–fluid interactions. Consequently, homogeneous flow is achieved even at low liquid velocities.

With particles larger than about 30 μm, a practical method of avoiding particle deposition in dilute-phase conveying is to increase the viscosity of the carrier fluid. This has the effect of reducing the terminal settling velocity of the particle, and as a result the critical velocity will also decrease. With a suitable medium, transportation may even be carried out in laminar or transitional flow, thus minimizing pipe erosion and energy requirement.

The majority of non-settling slurries also exhibit pronounced non-Newtonian flow behaviour even if the carrier fluid is initially Newtonian. The extent of the non-Newtonian behaviour is a complex function of particle–particle and particle–fluid interaction, solids concentration, particle and fluid properties, and operating conditions (Ayazi Shamlou 1984a,b). Unfortunately, there is no satisfactory theoretical formulation relating these factors to the non-Newtonian behaviour of the suspension.

An alternative engineering approach often adopted to study the flow properties of such a material is experimentally to obtain the 'flow curve' or 'shear diagram', which is a plot of shear stress against shear rate in the laminar region. This information is usually obtained in a laboratory viscometer. According to the shape of the flow curve for the particular material, an empirical correlation may be developed in order to model the fluid. Either the flow equation or the flow curve can then be used for

Hydraulic transport of particulate solids 139

full-scale design purposes. An alternative method commonly used in industry is to obtain pressure drop-flow rate data for the particular mixture using pilot-plant scale pipes. The information may then be used directly to design industrial-scale pipelines.

Rheological characteristics

For a simple Newtonian fluid, the viscosity is constant and independent of the magnitude of shear stress or shear rate. Consequently, the shear diagram for such a fluid is a straight line passing through the origin (Figure 6.4); the slope of the line gives the viscosity of the fluid.

For most non-settling slurries, the flow curves are not as simple as in the case of Newtonian materials. The most common types of non-Newtonian behaviour are:

(i) Time independent

These may be subdivided as follows (Figure 6.4):

(a) Shear thinning (pseudoplastic)

'Apparent viscosity', i.e. the ratio of the instantaneous shear stress to shear rate, falls as the shear rate increases. Examples of materials with such flow behaviour are many foodstuffs, inks, paints, paper pulp, emulsions, oils and greases.

(b) Shear thickening (dilatant)

'Apparent viscosity' increases as shear rate increases. Dilatant materials include some paints and inks, plastisols, some foodstuffs, and quicksand.

(c) Bingham plastic

With a yield stress which must be exceeded before flow commences. Typical examples include toothpaste, fresh cement/concrete, and suspensions of coal and many metal ores in water.

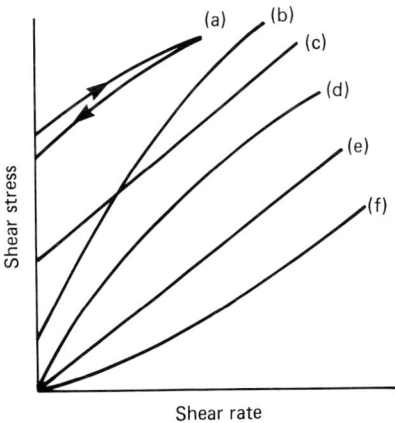

Figure 6.4 Flow curves for various non-Newtonian materials. (a) Thixotropic; (b) general non-Newtonian; (c) Bingham plastic; (d) pseudoplastic; (e) Newtonian; (f) dilatant.

(ii) Time dependent

These may be sub-divided into:

(a) Thixotropic

At any given shear rate 'apparent viscosity' decreases with time. It is rather like a pseudoplastic material but with a time effect. Examples include non-drip paints and inks, tomato ketchup, mayonnaise, drilling mud, and many other solid–liquid systems.

(b) Rheoplastic (antithixotropic)

At any given shear rate the 'apparent viscosity' increases with time. It is rather like a dilatant material but with a time effect, e.g. gypsum, paste, and bentonite suspension

(iii) Viscoelastic

These are materials whose properties show some resemblance to those of solids. These fluids exhibit elastic recovery from deformations which occur during flow. In spite of such effects, however, during steady state flow a viscoelastic fluid is often similar to a pseudoplastic material. Examples of viscoelastic materials include polymer melts and many polymer solutions, flour doughs, and asphalts.

In practice, combinations of different types of behaviour are also possible.

Rheological models

Many equations have been used to describe rheological data mathematically. Such mathematical models are required not only for describing and interpreting flow properties, but also for the hydraulic design of pipelines.

Ideally, a model for non-Newtonian flow should give an accurate fit of the experimental data and should have a minimum number of independent constants; the constants should be readily evaluated and should preferably have some physical basis.

For example, for dilitant and pseudoplastic materials with no yield stress, the logarithmic plot of shear stress against shear rate is often found to be a straight line over a limited range of shear rate. As a result, an empirical functional relationship known as the power law is widely used to characterize these materials. The relation may be written as:

$$\tau = K'\dot{\gamma}^{n'} \tag{6.24}$$

where $0 < n' < 1$ for pseudoplastic materials

and $n' < 1$ for dilatant materials

K' is a measure of the consistency of the fluid; the higher the value of K', the more viscous the fluid. n' is a measure of the degree of non-Newtonian

Hydraulic transport of particulate solids 141

behaviour with $n' = 1$ giving Newtonian fluid. The greater the departure from unity, the more non-Newtonian the character of the material.

The two rheological parameters K' and n' can be obtained from a number of measurements of flow through a capillary-tube viscometer or a pipe flow. From the measured values of the imposed pressure and resultant flow rate, the two parameters can be evaluated since the wall shear stress is given by the Hagen–Poiseuille relationship as:

$$\tau_w = \frac{\Delta PD}{4L} \tag{6.25}$$

and the Newtonian wall shear rate is given by:

$$\dot{\gamma} = \frac{8V}{D} = \frac{4Q}{\pi R^3} \tag{6.26}$$

Figure 6.5 is a log–log plot of τ_w against $\dot{\gamma}_w$ for various cement–water suspensions; the slope of the lines gives the value of n', and K' is the intercept at $\dot{\gamma}_w = 1$ on the τ_w axis.

The 'apparent viscosity', μ_a, of the material may be obtained from (Darby 1976):

$$\mu_a = \frac{\tau_w}{\dot{\gamma}_w} \left(\frac{4n}{3n + 1} \right) \tag{6.27}$$

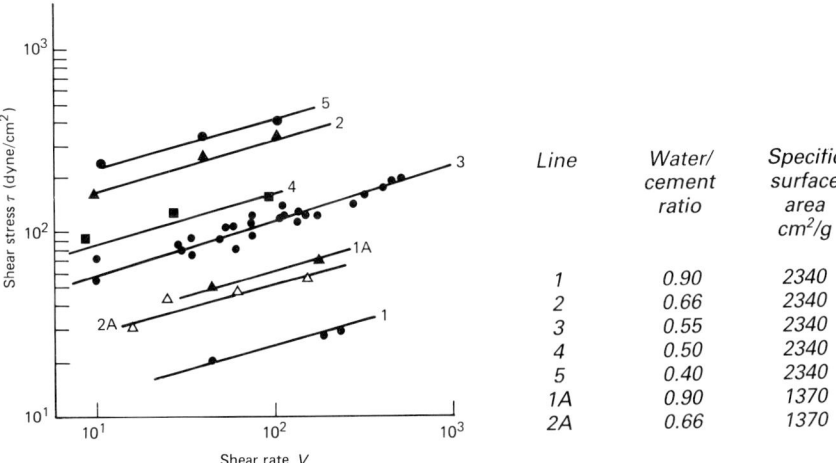

Line	Water/cement ratio	Specific surface area cm²/g
1	0.90	2340
2	0.66	2340
3	0.55	2340
4	0.50	2340
5	0.40	2340
1A	0.90	1370
2A	0.66	1370

Figure 6.5 Flow curves for various concentrations of cement–water slurries. (Data from Ayazi Shamlou, 1984a.)

142   Hydraulic transport of particulate solids

Other commonly used rheological equations for solid–liquid mixtures are:

$\tau = \tau_y + \mu_p \dot{\gamma} \quad \tau > \tau_y$ Bingham plastic
$\tau = \sqrt{\tau_y} + A\sqrt{\dot{\gamma}} \quad \tau > \tau_y$ Casson fluid
$\tau = \tau_y + K'\dot{\gamma}^{n'}$ general Bingham plastic

For more information on rheology and viscometry the reader is referred to such texts as Wilkinson (1960), and Darby (1976). The aim of this section is to use rheological properties in order to estimate the pressure drop for the hydraulic conveying of slurries.

Pressure drop for flow of non-settling slurries

Having established a suitable rheological model for the slurry, the next important stage in design is the estimation of the prevailing pressure drop during conveying. To do this, it is first necessary to establish whether the flow is laminar, transient or turbulent under the specified operating conditions. The criterion for the onset of turbulence adopted by many workers is by considering the conventional dimensionless Reynolds number, Re, defined for Newtonian fluids as $Re = \rho VD/\mu$.

For Newtonian fluids the transition from laminar to turbulent flow occurs rather suddenly over a narrow range of Reynolds number near 2000.

For non-Newtonian time-independent shear thinning materials obeying the power law relationship, Metzner and co-workers (1955, 1597, 1958) proposed the use of the generalized Reynolds number, defined as:

$$Re_{GEN} = \frac{D^{n'} V_m^{2-n'}}{K' 8^{n'-1}} \quad (6.28)$$

For flow in the laminar region, this definition of Reynolds number ensures that

$$f_f = \frac{16}{Re_{GEN}} \quad (6.29)$$

which is identical to the expression used for the estimation of energy losses for Newtonian fluids in laminar flow. f_f is the conventional Fanning friction factor defined as:

$$f_f = \left(\frac{D \Delta P}{4L}\right) \left(\frac{\rho V_m^2}{2}\right)^2 \quad (6.30)$$

Figure 6.6 shows the Fanning friction factor plotted against the generalized Reynolds number. The plot is obtained experimentally and may be used for the estimation of the pressure drop once the rheological parameters K' and n' have been evaluated.

In the laminar flow region, most available experimental measurements of pressure drop lie within $\pm 10\%$ of the Newtonian line, $f_f = 16/Re_{GEN}$ while in the turbulent flow regime:

$$f_f = fn(Re_{GEN}, n') \quad (6.31)$$

For turbulent flow of non-Newtonian materials in smooth pipes, Dodge and Metzner (1959) provided the following semi-empirical equation:

$$\frac{1}{\sqrt{f_{TS}}} = \frac{4}{n'^{0.75}} \log \left[Re_{GEN} f_{TS}^{1-n'/2} \right] - \frac{0.4}{n'^{1.2}} \quad (6.32)$$

which is similar to the von Karman expression for Newtonian liquids.

For turbulent flow of general Bingham plastic materials in partially rough-wall pipelines, Lazarus and Sive (1984) recommend the following alternative expression:

$$\frac{1}{\sqrt{f_{TPR}}} = -4 \log \left\{ \left[\frac{3.22}{Re_B f_{TPR}(1-\varphi)} \right]^{1.133} + \left(\frac{e}{3.35D} \right)^{1.018} \right\} \quad (6.33)$$

where $Re_B = \dfrac{\rho_m V_m D}{K_{eff}}$ with $K_{eff} = \left(1 + \dfrac{D\tau_y}{6K'V_m}\right)$

and $\varphi = \dfrac{\tau_y}{\tau_w}$

For smooth-wall flow the last term in { } in Equation 6.33 will be zero, thus giving;

$$\frac{1}{f_{TS}} = -4 \log \left[\frac{3.22}{Re_{GEN} f_f(1-\varphi)} \right]^{1.133} \quad (6.34)$$

while for fully rough-wall flow, the first term in { } in Equation 6.33 may be ignored. This reduces Equation 6.33 to:

$$\frac{1}{f_{TR}} = -4 \log \left[\frac{e}{3.35D} \right]^{1.018} \quad (6.35)$$

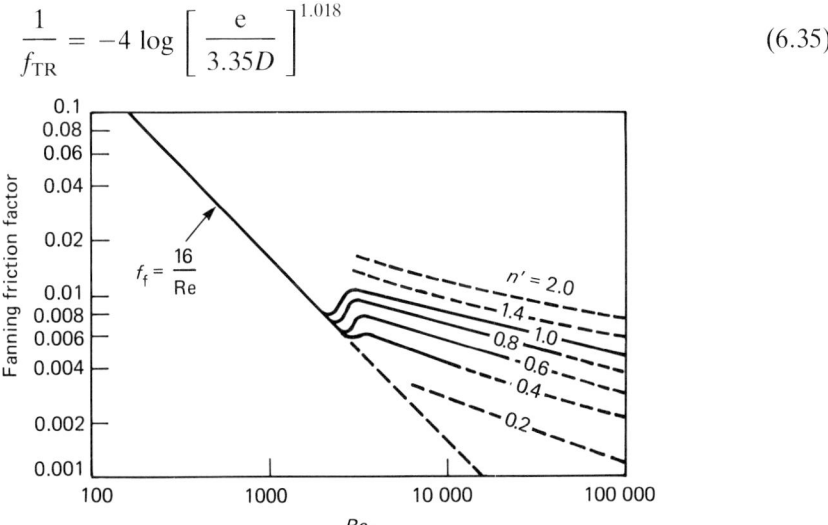

Figure 6.6 Friction factor-Reynolds number plot for non-Newtonian materials. (Data from Dodge and Metzner, 1959.)

Lazarus and Sive (1984) point out, however, that the constants in Equations 6.33–6.35 may have to be modified for different solid–liquid slurry systems.

One major shortcoming of the equations presented so far in this section is that the estimation of turbulent pressure drop is based on values of K' and n' that are obtained from laminar viscometer data. Bowen (1961) provided an alternative design procedure that overcomes this difficulty. Essentially the technique is based on measuring turbulent pressure-drop–flow rate data from small-scale equipment and using the results directly for industrial scale pipelines. The method does not depend upon any rheological model and is applicable to all flow regimes. It has the added advantage that wall slip, observed during small-scale experiments, may be accounted for during design.

The derivation of Bowen's design equations is based on a modification of the Blasius equation:

$$D^\alpha \tau_w = kV_m^\beta \tag{6.36}$$

which relates the wall shear stress, τ_w, to the mean pipe velocity. The constants α, k, and β are obtained experimentally from small-scale trials as outlined below:

(i) Plot the wall shear stress, τ_w, i.e. $(\Delta PD/4L)$, against the apparent shear rate $(8V_m/D)$ on log–log graph paper for all pipe diameters used. β is the slope of the lines, and should be constant provided that the relative roughness is the same in all the pipes used in the trials.
(ii) Replot the data in the form $(\Delta P/V_m^\beta)$ against pipe diameter D on log–log graph paper. Data for all pipe diameters should coincide. The intercept on this plot is equal to $4k$ and the slope of the line is given by $-(1 + \alpha)$.

The experimentally established values of α and β allow the pressure drop to be estimated for full-size pipe diameters.

α and β are affected by the roughness of the pipe as well as by the properties of the fluid. Lord *et al.* (1967), however, have reported a mean value of $\alpha = 0.2$ for various suspensions, while Harris (1968) claims excellent agreement between experimentally observed pressure drops and predicted values using Bowen's method.

Kenchington (1972) found that Bowen's approach could correlate data from small-scale experiments to within ± 7%, but results from scale-up studies were far less satisfactory, with 90% confidence limits falling in the range 65% below and 165% above the best fit line. Carleton and Cheng (1977) point out that Kenchington's conclusions are inevitable because of the variable nature of most industrial slurries. These workers suggest that if a high degree of accuracy is needed, then experimental trials on large-scale equipment are essential.

Pressure drop in pipe fittings for flow of non-settling slurries

There has been little systematic investigation of the flow of non-settling slurries through bends, valves and fittings. Such information is obviously needed for a proper design of any pipeline transportation network (Turian et al., 1983).

In the absence of any fundamental work, and as a rule of thumb, an allowance of 12 m should be deducted from the maximum horizontal distance for every right angle bend, and 2.5 m for every 0.3 m of head.

Cheng and Higman (1969) suggest, as a first approximation, that under fully developed turbulent flow conditions, the pressure drops in pipe fittings are practically independent of the non-Newtonian properties. At low flow rates, however, the non-Newtonian properties are significant and could lead to practical problems if proper precautions are not taken at the design stage.

In summary, a large number of correlations and procedures have been developed for the design of hydraulic transport systems for both settling suspensions and non-settling slurries. The vast majority of these correlations are empirical in nature and as such are unsatisfactory outside the range of experimental conditions for which they were developed. For this reason, design should be based on experimental trials on either the actual product to be conveyed or on a material with characteristics similar to it. Moreover, at the present stage of development, great uncertainties still exist with regard to the scale-up of data obtained from flow experiments on pilot-plant pipeline. Consequently, the most reliable method of design is to obtain the relevant parameters from experimental tests, using equipment on a scale as close to the full size as is practicable.

Equipment components

Pumps for hydraulic transportation

Centrifugal and positive-displacement pumps are the two main types of pumping equipment for the hydraulic transport of non-settling suspensions and slurries.

Centrifugal pumps are generally suitable for low-pressure applications, although with multiple pumps in series discharge pressures of up to 600 p.s.i. may be achieved. With fine abrasive slurries and low-pressure applications, rubber-lining of the pump is generally sufficient to eliminate wear, while with large particles wear-resistant alloys, e.g. nickel hardened, are usually employed (Aude et al., 1971).

With pumping pressures above 600 p.s.i. and for very abrasive materials, positive displacement, plunger-type pumps are used. For less abrasive suspensions, single- and double-acting piston pumps can also be used.

From a design point of view, a slurry pump (centrifual or positive-displacement) looks very much like the standard liquid version, particularly from the outside. Indeed, in prinicple, the design philosophy is the same in both cases. This basically involves matching the required system curve (plot of the total head loss versus capacity for the pipe network) against the pump characteristic curve (also a plot of total head developed versus flow rate).

The main difference between a slurry pump and a standard fluid pump lies in the design and construction of the inner components of the unit, and it is this which makes the slurry pump a highly specialized device: the design of all thicknesses, protrusions within the pump, flow passages, material of construction and linings become critically important with slurry pumps (Dalstad, 1977; Aude et al., 1971).

With highly abrasive slurries it is possible to isolate the mixture from the pumping device. The 'lock hopper' (Figure 6.7) is one such system which allows the use of conventional multistage water pumps to transport slurries (Aude et al., 1971).

Valves

In designing and selecting values for slurry and suspension applications, the most important consideration is the abrasive nature of the material. To minimize wear, the valve should ideally provide full opening with no dead zones in which the particles could get trapped. For low-pressure applications rubber-to-rubber or rubber-to-metal sealing is usually adequate while for high-pressure transportation lubricated plug or ball valves are employed (Aude et al., 1971) even though they are not entirely satisfactory.

Instrumentation

The measurement of control variables, such as solids concentration and suspension flow rate, also require special attention, since many conventional meausring devices used for liquids are unsuitable for suspension

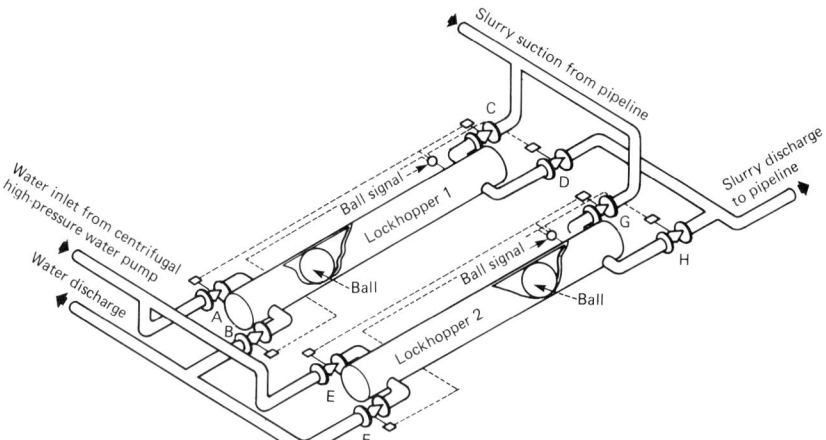

Figure 6.7 Lockhopper system allows the use of liquid centrifugal pumps by isolating the abrasive slurry. *Lockhopper 1* is shown discharging slurry to the pipeline. Valves B and C are closed, whereas A and D are open. Water enters through Valve E, forcing the ball against the slurry which in turn is forced out of the hopper through Valve D, and thus into the pipeline. The ball will move to the right until it trips the signal. The relay from this signal causes A and D to close, while B and C are opened. *Lockhopper 2*, shown in the filling stage, is receiving slurry through Valve G, and discharging water through Valve F. The hopper will continue filling with slurry until the ball trips a signal (not shown). The signal will cause F and G to close, and E to H to open, thus enabling the now-full hopper to discharge slurry into the pipeline.

applications. For example, most fluid flow meters such as rotating blades, vortex-shedding, venturi and orifice plate meters, are inappropriate because of clogging and abrasion. Non-intrusive flowmeters, e.g. electromagnetic and ultrasonic meters, are preferable for slurry applications, since they create no restriction to flow: such meters are however generally more expensive than the mechanical and the differential pressure types, and require care in installation.

Suspension density, that is the per cent of suspended solids in the liquid, is another important variable that often needs to be measured for control purposes. Notice that suspension density is not quite the same as specific gravity as it does not measure either the clear liquid density or the dissolved solids.

Suspension density may be measured on line using either the U-tube densitometer or the gamma-ray radiation density gauge (Liptak, 1967): radiation meters are usually preferred since they provide no obstruction to flow, but, would require frequent calibration (Aude et al., 1971).

References

AUDE, T. C., COWPER, N. T., THOMPSON, T. L. and WASP, E. J. (1971) (June 28), Chem. Eng. 74–90
AYAZI SHAMLOU, P. (1984a). IChemE, *Fluid Mixing II*, Symp. Series No. 89, 3–5 April, 15–27
AYAZI SHAMLOU, P. (1984b). *The Chemical Engineer*, No. **403**, 31–34
BABCOCK, H. A. (1971). In *Advances in Solid–Liquid Flow in Pipes and its Application*, Zandi, I. (Ed.). Pergamon Press, New York
BOWEN, R. LeB. (1961). *Chem. Engng.* 24 July 143–150
BROOKES, D. A. and DODWELL, C. H. (1984). Hydrotransport 9, Rome, Italy, 17–19 Oct. 33–47
CARLETON, R. P. and CHENG, D. C.-H. (1977). *Chem. Engng.* April 25, 95–100
CHENG, D. C.-H. and HIGMAN, R. W. (1969). *Proc. Engng.* (July) 19–24
CHHABRA, R. P. and RICHARDSON, J. F. (1983). *Chem. Eng. Res. Dev.*, **61**, pp. 313–317
CONDOLIOS, E. and CHAPUS, E. E. (1963a). *Chem. Engng.* (8 July) 131–138
CONDOLIOS, E. and CHAPUS, E. E. (1963b). *Chem. Engng.* (24 July) 93–98
DALSTAD, J. I. (1977). *Chem. Engng.* (25 April) 101–106
DARBY, R. (1976). *Viscoelastic Fluids, An Introduction to Their Properties and Behaviour*, Vol. 9, Marcel Dekker, New York
DODGE, D. W. and METZNER, A. B. (1959). *AIChE J.*, **5**, 189–207
DURAND, R. (1953). *Proc. Minn. Int. Hyd. Conv.*, Sept. 1953
DURAND, R. and CONDOLIOS, E. (1952). Soc. Hyd. de France, Grenoble
FADDICK, R. R. (1982). Hydrotransport 8, Johannesburg, South Africa, Aug. 25–27, 37–49
HAAS, D. B., HUSBAND, W. H. W., SCHRIEK, W. and SMITH, L. G. (1973). Saskatchewan Reseach Council Reports 1–8, Aug
HARNBY, N., EDWARDS, M. F. and NIENOW, A. W. (Eds), (1985). *Mixing in the Process Industries*. Butterworths, London
HARRIS, R. W. (1968). *Rheol. Acta*, **7**, 228
HAYDEN, J. W. and STELSON, T. E. (1971). In *Advances in Solid–Liquid Flow in Pipes and its Application*, Zandi, I (Ed), Pergamon Press, New York, 149–164
KENCHINGTON, J. M. (1972). Hydrotransport 2, University of Warwick, Sept., Paper C4
LAZARUS, J. H. (1982). Hydrotransport 8, Johannesburg, South Africa, Aug. 25–27, 123–133
LAZARUS, J. H. and SIVE, A. W. (1984). Hydrotransport 9, Rome, Italy, Oct. 17–19, 207–227
LIPTAK, B. G. (1967). Chem. Engng (13 Feb), 151–158
LORD, D. L., HUSLEY, B. W. and MELTON, L. L. (1967). *Soc. Pet. Eng. J.*, **7**, 252
METZNER, A. B. and FRIEND, W. L. (1958). *AIChE J.*, **4**, 393–402
METZNER, A. B. and OTTO, R. E. (1957). *AIChE J.*, **3**, 3–11

METZNER, A. B. and REED, J. C. (1955). *AIChE J.*, **1**, 434–440
NEWITT, D. M., RICHARDSON, J. F., ABBOTT, M. and TURTLE, R. B. (1955). *Trans. Instn. Chem. Engrs*, **33**, 93–113
NEWITT, D. M., RICHARDSON, J. F. and GLIDDON, B. J. (1961). *Trans. Instn. Chem. Engrs*, **39**, 93–100
ORR, C. JR. (1966). *Particulate Technology*, Macmillan, New York
STEPANOFF, A. J. (1969). *Gravity Flow of Bulk Solids and Transportation of Solids in Suspension.* John Wiley, New York
STREAT, M. (1982). Hydrotransport 9, Johannesburg, South Africa, Aug. 25–27, 111–123
SVAROVSKY, L. (Ed) (1981). *Solid-Liquid Separation*, Butterworth, London
TURIAN, R. M., HSU, F. L. and SAMI SELIM, M. (1983). *Particulate Science and Technol.*, **1**, 365–392
TURIAN, R. M. and YUAN, T. (1977). *AIChE J.*, **23**, 232–243
WHITE, J. F. C. and SEAL, M. E. J. (1982). Hydrotransport 8, Johannesburg, South Africa, Aug. 25–27, 63–75
WILKINSON, W. L. (1960). *Non-Newtonian Fluids – Fluid Mechanics, Mixing and Heat Transfer.* Pergamon Press, New York
ZANDI, I. and GOVATOS, G. (1967). J. Hyd. Div., Proc. ASCE, HY3

Symbols

A	constant in Equation 6.1
C_d	single particle drag coefficient $\left[4/3 \dfrac{gdp(s-1)}{U_t} \right]$
$C_{d_{wm}}$	weighted mean drag coefficient (Equation 6.5)
C_v	solids concentration (volume fraction)
C_w	solids concentration (weight fraction)
D	pipe diameter
d_p	particle diameter
e	mean size of pipe roughness
f_f	Fanning friction factor for flow of fluid alone
f_T	turbulent friction factor for the flow of suspension
F_{TS}	turbulent friction factor for flow in smooth-wall tube
$f_{T_{PR}}$	turbulent friction factor for flow in rough-wall tube
f	turbulent friction factor for flow in partially rough-wall tube
F_L	constant in Equation 6.3
g	acceleration due to gravity
i_T	total head loss per unit length of pipe due to flow of suspension
i_p	head loss per unit length of pipe due to flow of fluid alone
i_f	head loss per unit length of pipe due to the presence of solids in the suspension
K'	consistency index
L	pipe length
m	constant in Equation 6.1
n'	flow behaviour index
N	parameter defined by Equation 6.11
ΔP	pressure drop
R	pipe radius
Re	Newtonian Reynolds number $(\rho VD/\mu)$
Re_{GEN}	non-Newtonian Reynolds number given by Equation 6.28
Re_B	Bingham plastic Reynolds number $(\rho V_m D/K')$

s	ratio of particle to fluid density (ρ_p/ρ)
U_t	single-particle terminal settling velocity
V	fluid velocity
v_c	critical deposit velocity
V_m	mean suspension velocity
$V_{ho \to ht}$	transition velocity from homogeneous to heterogeneous flow
$V_{ht \to sl}$	transition velocity from heterogeneous to sliding-bed flow
$V_{sa \to ho}$	transition velocity from saltation to homogeneous flow
$V_{st \to ht}$	transition velocity from stationary to heterogeneous flow
$V_{st \to ho}$	transition velocity from stationary to homogeneous flow
α	constant in Equation 6.36
β	constant in Equation 6.36
$\dot{\gamma}$	shear rate
$\dot{\gamma}_w$	wall shear rate
μ	Newtonian viscosity
μ_a	apparent viscosity of non-Newtonian fluid
μ_p	plastic viscosity
ρ	fluid density
ρ_p	particle density
ρ_m	mixture density
φ	ratio of yield stress to wall shear stress (τ_y/τ_w)
ψ	parameter defined by Equation 6.15

Example 1. Settling Suspensions

Coal particles with a mean size of 500 μm are to be conveyed hydraulically in a 0.15 m diameter pipe. If the volume fraction of solid particles is kept at 0.1, determine the critical mixture velocity needed to avoid particle deposition. If the actual operating velocity is 12% above this critical value, calculate the total throughput, the hydraulic gradient and the energy loss during transportation. Other relevant properties are:

ρ_f = 1300 kg/m³
ρ_p = 2000 kg/m³
μ_f = 0.001 kg/ms
pipe roughness, e/D = .01
and $f_f = 3.2 - 2.5 \ln (e/D)$

Solution

To obtain the critical velocity, the particle drag coefficient must first be calculated. This is done using standard drag charts (Figures 6.8 and 6.9). Thus:

$$\frac{1}{2} C_d Re^2 = \frac{4 d_p^3 \rho_f (\rho_p - \rho_f) g}{3 \mu_f^2}$$

$$= \frac{4 \times (500 \times 10^{-6}) \times 1300 \times (2000 - 1300) \times 9.81}{3 \times (0.001)^2}$$

$$= 1487$$

Thus from Figures 6.8 and 6.9

$Re_p = 25 \qquad C_d = 0.5$

The critical deposit velocity may now be obtained using Equation 6.3:

$$V_c = F_L \sqrt{2gD(s - 1)}$$

where

$$F_L = \frac{2.43(C_v)^{1/3}}{\sqrt[4]{C_d}}$$

Thus

$$F_L = \frac{2.43 \times (0.1)^{1/3}}{\sqrt[4]{(0.5)}} = 1.34$$

and

$$V_c = 1.34 \times 2 \times 9.81 \times \left[\frac{2}{1.3} - 1\right]$$

$$= 1.7 \text{ m/s}$$

The operating mixture velocity, V_m, is 12% higher than this. Thus

$V_m = 1.89$ m/s

V_m is now used to calculate the total throughput, Q:

$$Q = \frac{V_m \pi D^2}{4} = 0.033 \text{ m}^3/\text{s}$$

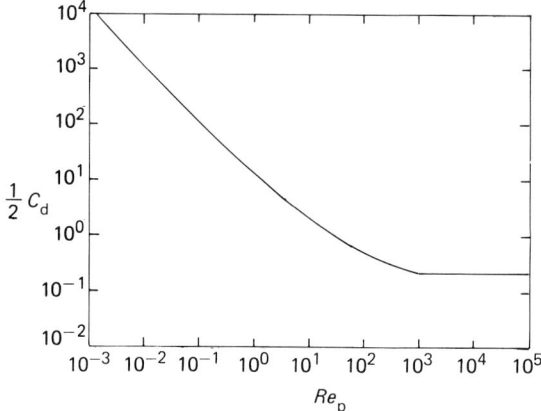

Figure 6.8 $1/2C_d$ vs Re_p for spherical particles.

Hydraulic transport of particulate solids

To calculate the hydraulic gradient it is first necessary to establish the flow regime using Equations 6.9 and 6.10. Thus:

$$V_{ho \to ht} = [1800g\, D\, U_t]^{1/3}$$

where U_t is the free-fall velocity and may be obtained using Re_p: Thus

$$Re = \frac{\rho_f U_t d_p}{\mu_f} = 25$$

$$U_t = \frac{25 \times 0.001}{(500 \times 10^{-6}) \times 1300} = 0.04 \text{ m/s}$$

Therefore

$$V_{ho \to ht} = [1880 \times 9.81 \times 0.15 \times 0.04]$$
$$= 8.8 \text{ m/s}$$

while,

$$V_{ht \to sl} = 17[U_t]$$
$$= 0.68 \text{ m/s}$$

with $V_{ho \to ht} < V_m < V_{ht \to sl}$ the flow is heterogeneous and thus Equation 6.7 may be used to calculate the total hydraulic gradient:

$$\frac{i_T + i_f}{C_v i_f} = 1100(s-1)\left(\frac{gD}{V^2}\right)\left(\frac{U_t}{V_m}\right)$$

$$\frac{i_T - i_f}{0.1\, i_f} = 1100\left[\frac{2}{1.3} - 1\right]\left(\frac{9.81 \times 0.15}{1.89^2}\right)\left(\frac{0.04}{1.89}\right)$$

$$= 5$$

thus $i_T = 1.5\, i_f$

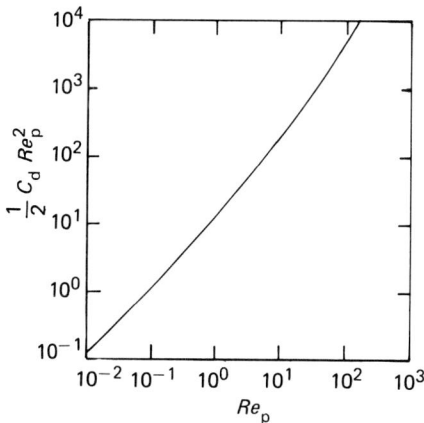

Figure 6.9 $1/2 C_d Re_p^2$ vs Re for spherical particles.

152 Hydraulic transport of particulate solids

where i_f is the hydraulic gradient for the flow of fluid alone and is obtained using Equation 6.2, that is:

$$i_f = 2f_f \frac{V_m^2}{gD}$$

with $f_f = 3.2 - \ln(e/D)$
$= 3.2 - \ln(0.01)$
$= 0.005$

$$\therefore \quad i_f = \frac{2 \times 0.005 \times 1.89^2}{9.81 \times 0.15} = \frac{0.025 \text{ m water/m}}{(310 \text{ N/m}^2/\text{m})}$$

thus $i_T = 0.0378$ m water/m
$(\equiv 482 \text{ N/m}^2/\text{m})$

also, the total energy loss, $i_T Q = 16$ W/m

It is perhaps worth noting that the criterion proposed by Zandi and Govatos, Equation 6.16, will give the following results for the critical deposit velocity:

$$V_c^2 = 40 \frac{C_v gD(s-1)}{\sqrt{C_d}}$$

$$= 40 \times \frac{0.1 \times 0.15 \times 9.81 \times (1.54 - 1)}{\sqrt{0.5}}$$

thus $V_c = 2.1$ m/s and $V_m = 2.35$ m/s

Furthermore, from Equation 6.15,

$$\psi = \frac{V_m^2}{gD(s-1)} \sqrt{C_d}$$

$$= \frac{2.35^2}{9.81 \times 0.15 \times (1.54-1)} \times \sqrt{0.5}$$

$$= 5$$

with $\psi < 10$, Equation 6.13 is used to estimate the total hydraulic gradient, that is:

$$\frac{i_T - i_f}{C_v i_f} = 280 \, \psi^{-1.93}$$

giving $i_T = 0.057$ m water/m $(\equiv 732 \text{ N/m}^2/\text{m})$

which compares well with the value predicted by Equation 6.1 with $m = 1.5$ and $A = 85$, that is:

$$\frac{i_T - i_f}{C_v i_f} = 85 \left[\frac{gD(s-1)}{V} \frac{1}{\sqrt{C_d}} \right]^{1.5}$$

therefore:

$$\frac{i_T - 0.025}{0.0025} = 85 \left[\frac{9.81 \times 0.15 \times 0.54}{1.89^2} \frac{1}{\sqrt{0.5}} \right]^{1.5}$$

thus:

$i_T = 795$ N/m²/m

Example 2. Non-settling slurries

Figure 6.10 shows pressure drop–velocity data (plotted in terms of shear stress $\Delta pD/4L$ vs shear rate, $8V_m/D$) for the turbulent flow of a cement slurry in 0.0127 m, 0.0190 m and 0.0254 m diameter pipes. Using Bowen's method, estimate the pressure drop for the same slurry flowing in a 0.2 m diameter pipeline at a rate of 0.05 m³/s.

Solution

To use Bowen's method, the magnitude of the constants k, α, and β must first be evaluated:

(i) β is obtained directly from the slope of the plots in Figure 6.9. The lines in Figure 6.9 are all parallel having the same slope which is:

∴ $\beta = 0.9$ (mean value)

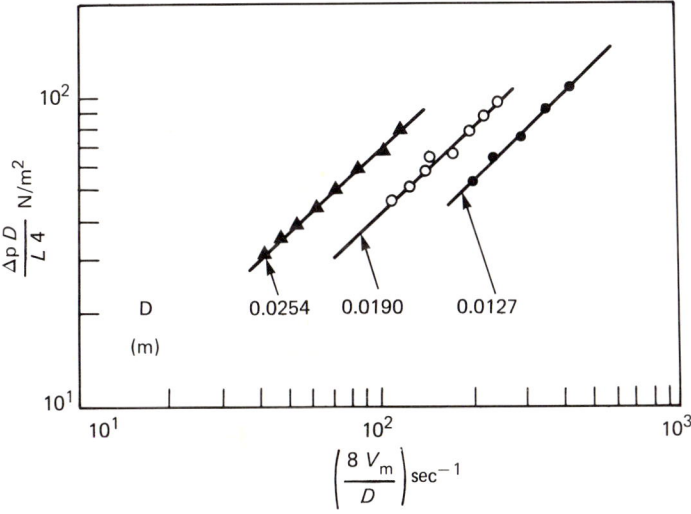

Figure 6.10 Shear stress vs shear rate for various pipe diamters.

154 Hydraulic transport of particulate solids

(ii) The table below shows the values of V_m, $\Delta p/LV_m^\beta$ and D

D (m)	V_m m/s	$\dfrac{\Delta p}{LV_m^\beta}$	D (m)	V_m m/s	$\dfrac{\Delta p}{LV_m^\beta}$	D (m)	V_m m/s	$\dfrac{\Delta p}{LV_m^\beta}$
.0127	0.6	63403	.019	0.45	34723	.0254	0.25	31482
	0.5	65087		0.38	35568		0.22	29252
	0.43	63348		0.29	39063		0.18	28250
	0.38	63354		0.26	35921		0.13	29809
mean		63798			36319			29697

Figure 6.11 shows the plot of mean $\Delta p/LV_m^\beta$ against D from which:

$\alpha = 0.11$ and $k = 158$

Thus Bowen's Equation 6.36, that is:

$$\tau_w D^\alpha = kV_m^\beta$$

$$\left(\dfrac{\Delta p D}{L4}\right) D^\alpha = kV_m^\beta$$

may be rewritten as:

$$V_m = \left[\dfrac{(D)^{1+\alpha}\Delta P}{4kL}\right]^{1/\beta}$$

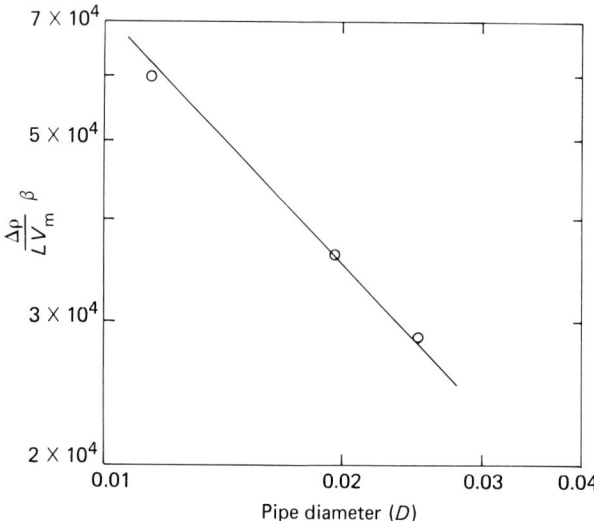

Figure 6.11 $\Delta p/LV^\beta$ vs pipe diameter.

thus:

$$V_m = \left[\frac{D^{1.11}}{632} \frac{\Delta p}{L} \right]^{1/0.9}$$

or

$$\frac{\Delta p}{L} = {}^{1/0.9}\sqrt{V_m} \frac{632}{D^{1.11}}$$

therefore for $D = 0.2$ m and $Q = 0.05$ m³/s

$$V_m = \frac{Q}{\pi D^2/4} = \frac{0.05}{\pi(0.2^2/4)} = 1.6 \text{ m/s}$$

$$\therefore \frac{\Delta p}{L} = \frac{632}{(0.2)^{1.11}} {}^{1/0.9}\sqrt{1.6}$$

$$= 4122 \text{ N/m}^2/\text{m}$$

Chapter 7
Mechanical conveyors

Introduction

Mechanical conveyors are the workhorses of most bulk solids handling plants. They are used not only for the transportation of bulk solids from one location in the plant to another but also for feeding, discharging, metering and proportioning such materials to and from bulk solids storage silos, and other solids handling and processing equipment.

There are many types of mechanical conveying and elevating equipment. A convenient classification of some of the more important conveyors is as follows:

(i) belt conveyors;
(ii) screw conveyors/elevators;
(iii) bucket elevators;
(vi) chain conveyors/elevators (en masse conveyors and elevators).

In practice, very often for a given conveying duty a range of possibilities exists as illustrated in Figure 7.1. The choice of conveyor depends largely upon the specified capacity, conveying distance and configuration (whether horizontal, vertical or inclined), bulk solids and individual particle properties, particularly bulk density, maximum particle size, abrasiveness, flowability, toxicity, corrosiveness, explosibility and temperature (Colijn, 1985). For example, bulk solids with a temperature above about 120° C can not normally be transported by belt conveyors, but provided the conveying distance is below about 100 m and the bulk material is not friable, then screw conveyors may be used effectively. For bulk solids with temperatures below approximately 80° C belt conveyors have been used successfully to cover horizontal distances of over 300 m by a single belt. However, belt conveyors cannot be used for vertical lifts; for such duties, bucket or screw elevators are often more appropriate.

Mechanical conveyors are also used widely as feeders to damp out fluctuations and control flow rates of bulk solids from storage silos and other solids handling and processing equipment. As far as storage silos are concerned, the feeder should be designed to give uniform flow by ensuring that the silo outlet is fully activated. With circular and square openings, almost any type of feeder may be used, provided that a vertical spout with a length approximately equal to one outlet diameter is used to connect the feeder to the storage silo (Johanson, 1969). If the specified flow rate results in a feeder that is smaller than the silo outlet, then a bin activator (Figure 7.2) or some other form of pre-feeder device is generally recommended (Doeksen, 1973).

With slotted openings, in order to obtain uniform flow, the design of the feeder must allow for increased capacity in the direction of flow. Johanson (1969) has highlighted many of these problems and has also provided practical solutions to them. Some of these are shown in Figure 7.3.

Screw conveyors and elevators

In a screw conveyor, transportation of bulk material is achieved by the rotation of a helical screw thus moving the material axially in a trough or tube (Figures 7.4, 7.5). This method of transportation is extremely versatile and may be used to convey, elevate, feed and discharge bulk solids.

Power requirement and discharge capacity are affected strongly by the conveying distance, the angle of inclination, and the geometrical configuration of the screw. The maximum size of the particles should be smaller than

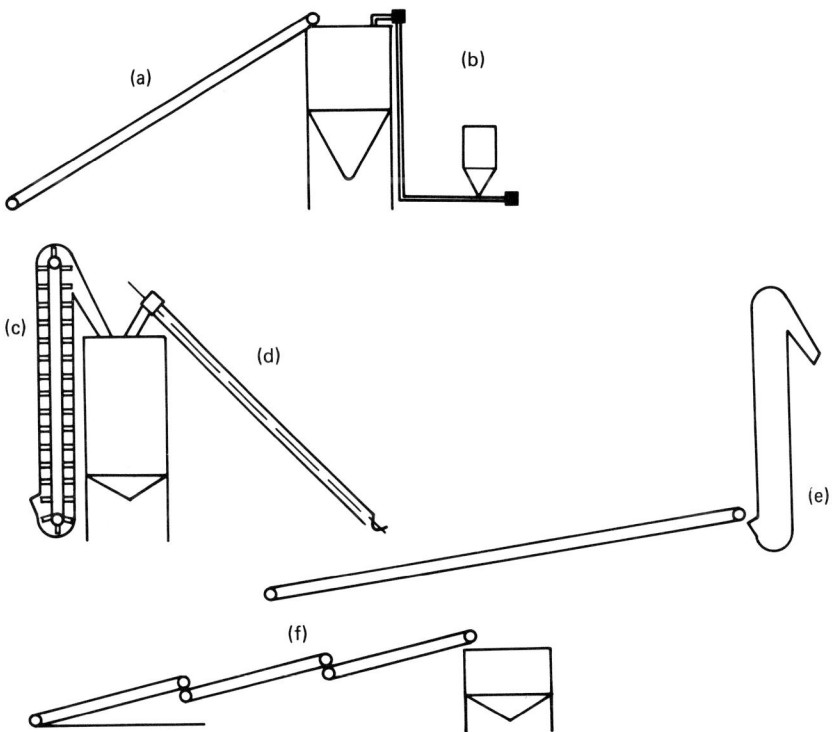

Figure 7.1 Typical mechanical conveyors. (a) Belt elevator; (b) screw elevator; (c) bucket elevator; (d) screw conveyor; (e) belt conveyor and bucket elevator; (f) multiple belt conveyors.

158 Mechanical conveyers

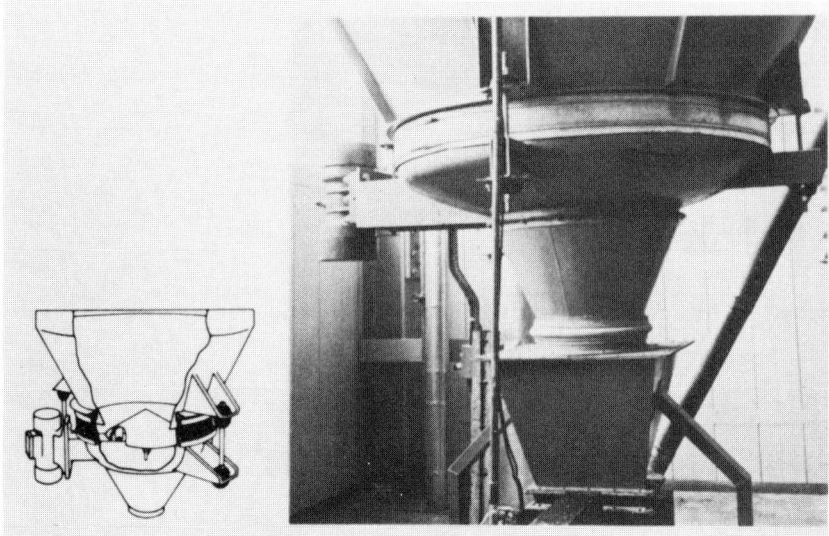

Figure 7.2 Grain silo bin activator.

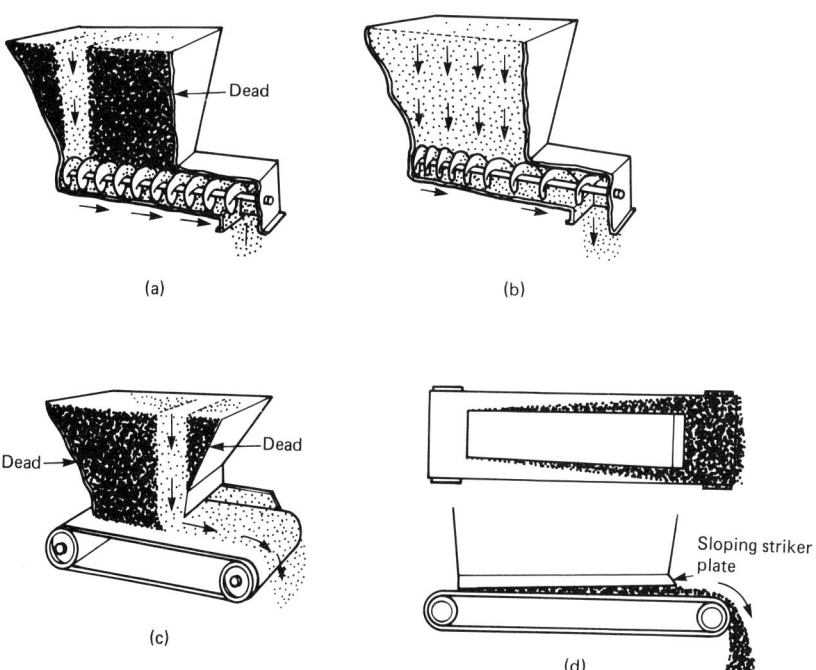

Figure 7.3 Screw and belt feeders. (a) Constant-pitch screw; (b) variable-pitch screw; (c) parallel-slot belt feeder; (d) tapered-slot belt feeder.

Mechanical conveyers 159

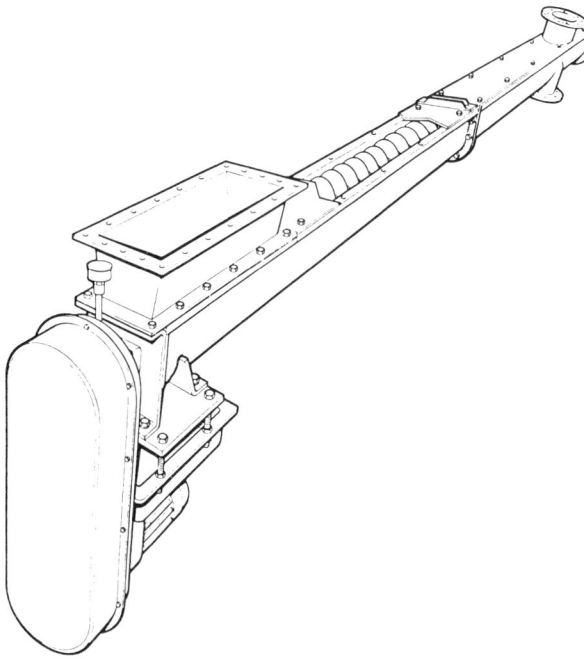

Figure 7.4 Screw feeder/discharger.

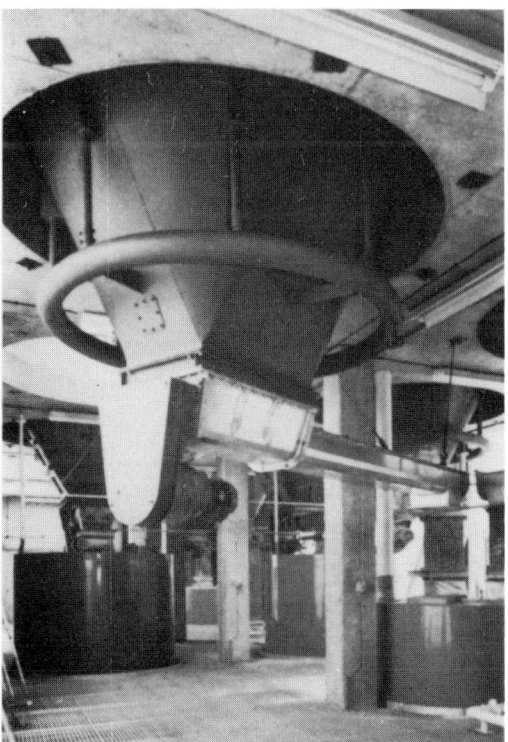

Figure 7.5 Screw feeder/discharger attached to bulk powder storage silo.

160 Mechanical conveyers

the gap between the rotating blades and the inner walls of the casing; this gap should be as small as practicable in order to increase conveying efficiency.

The casing can be a trough with a square, rectangular, U-shaped or flared cross-section. The top of the trough may be open or, alternatively, fitted with a quick-release cover to facilitate cleaning. For angles of inclination greater than about 30°, a circular casing is usually recommended.

Outlets and inlets may be shaped and fitted along the case as required to suit the application; the various designs available commercially include motorized, pneumatic, rack and pinion, and hand-operated outlet slides, and plain, flanged or spigotted inlets.

A large range of flight designs is available to cover almost all conceivable applications (Figures 7.6, 7.7). For most conveying applications, a single screw with a continuous flighting and a constant pitch-to-diameter ratio of unity is the standard geometry used. For very easy-flowing materials a pitch equal to $1^1/_2$ times the screw diameter may be sufficient to move the bulk material effectively, while for vertical applications and for screw feeders the pitch to diameter ratio is usually $^2/_3$ to $^1/_2$.

For uniform discharge of fine (floodable) powders over the full length of the feed opening, a continuous screw with a constant diameter but increasing pitch is often more appropriate than a screw with a constant pitch (Figure 7.3, 7.6): for some applications, e.g. feeding lumpy materials, the screw diameter may also be tapered, usually from a pitch to diameter ratio of $^2/_3$ to 1.0 (Figures 7.3, 7.6). When the screw conveyor is used as a feeder, the length to diameter ratio is usually limited to 6 in order to ensure uniform feeding. This limitation makes screw conveyors unsuitable as feeders for extremely long slotted openings (Johanson, 1969).

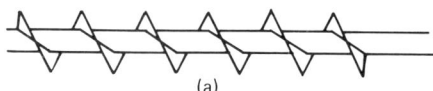

(a)

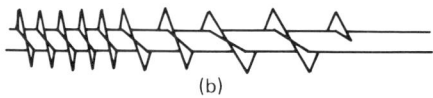

(b)

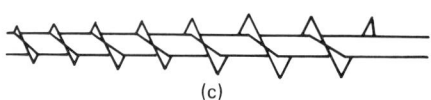

(c)

Figure 7.6 Typical screw flights. (a) Constant-pitch; (b) variable-pitch; (c) variable-pitch, tapered.

Mechanical conveyors 161

Other flight designs are available for specialized duties; these include wire screw, double-flight screw, single- and double-flight ribbons, single cut-flight and cut and folded flight screws with standard pitch.

The energy requirement and discharge rate of screw conveyors are affected by the speed of rotation, screw size, pitch and length, degree of fill and angle of inclination from horizontal. Screw power requirement and flow rate are also affected critically by bulk solids and particle properties such as size, size distribution, bulk density, angle of internal friction and angle of wall friction between the particles and the wall of the housing and the rotating blades (Carleton *et al.*, 1969). In view of the large number of variables involved, no satisfactory theoretical expression has yet emerged for the estimation of the power input and the rate of discharge for industrial-scale installations. In practice, therefore, design is usually based upon previous experience, manufacturers' charts and/or test results from small-scale experiments using the actual powder which is to be transported in the full-scale unit. In this connection, some available experimental evidence suggests that large-diameter screws are relatively more efficient than smaller ones and thus design based on small-scale tests will tend to be conservative.

Belt conveyors

Belt conveyors are among the most versatile and widely used mechanical conveyors. In the process industry, they are used for transporting, proportioning, feeding, discharging and metering bulk solid materials. This

Figure 7.7 Typical screw flights.

method of transportation is suitable to almost any type of material from dry, free-flowing solids to wet, sticky and lumpy powders provided the temperature of the material is not above 120° C.

The belt conveyor is essentially an endless belt on which the solids are transported. The belt is friction driven at one end and is carried on an idling drum at the other end. With normal belts, materials may be conveyed at angles of inclination of up to 22°, while non-slip devices mounted on the surface of the belt may be employed to increase the inclination to about 45°. High-angle belt conveyors with belt widths of 2 m, lift heights of about 94 m, angles of inclination of 35.5°, belt speeds of 2.7 m/s and conveying capacities of 4400 t/hr have been designed and installed successfully (Anon, 1985).

Belt materials vary widely, ranging from polyester covered with rubber and plain or coated canvas, to woven wire or steel ribbon. The belt is usually flat, but for transportation bulk solid materials, it is normally troughed, using idling rollers or slides. To prevent the belt slipping on the pulleys, it is necessary to place a certain initial tension in the belt. The actual tension acting on the belt during operation is larger than the initial value by an amount which is equivalent to the power input to the belt. The ratio of the total tension in the belt to the initial tension is known as the transmission ratio and has a value ranging from about 1.9 for a simple single bare drive to about 1.1 for a tandem-logged drive.

The total tension per metre of belt width together with the working stress per metre per ply for the type of belt may be used to evaluate the number of plies needed for any particular installation: if the belt is too thin it will tend to sag between the idlers, while if it is too thick it will not trough properly.

Troughing rollers require periodic repair and replacement often due to excessive wear. An alternative design that eliminates the need for troughing rollers is the air-supported belt conveyor (Read, 1985). In this design (Figure 7.8), the belt travels on a cushion of air in a trough that forms the

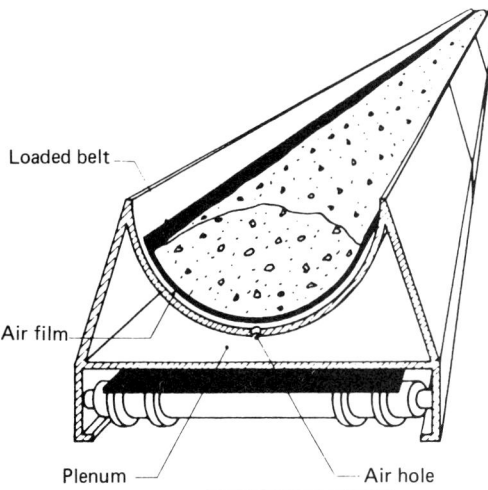

Figure 7.8 Air-support belt conveyor.

upper surface of a plenum. Small holes at intervals allow air to flow to create a frictionless layer of air between the trough and the underside of the belt. Air-supported belt conveyors have been used successfully to convey easy-to-handle materials such as grains and wood chips at conveying capacities below 200 t/hr. Read (1985) points out that without the troughing rollers and their resistance, the power consumed by the air-belt conveyor is reduced, and this together with the reduction in the cost of maintaining troughing rollers make this type of conveyor a very attractive alternative to conventional belt conveyors. Read (1985) also claims that air-supported belt conveyors could be designed to convey bulk solids at rates of well over 3000 t/hr, and that such units can also convey cohesive powders.

The bulk material on the belt conveyor is discharged over the drive pulley on other conveyors or, alternatively, into storage bins, mixers and other process equipment; special designs are available which allow multipoint discharge.

Belt conveyors are also used extensively as feeders. Figures 7.3 and 7.14 show some typical arrangements. Belt feeders are effective for solids below plus 6-in, and are particularly suited to long slotted hopper outlets; this type of feeder has been used successfully under slot openings as long as 30 m with a constant slot width of 0.2 m (Johanson, 1969).

The complete design of belt conveyors requires the specification of belt width, speed, capacity and power requirement. The capacity of the belt conveyor depends upon the speed, shape and width of the belt, method of feeding and solids properties, e.g. bulk density and particle size. The power input to the belt conveyor is made up of several components: (i) power needed to move the load, (ii) power necessary to move the belt itself, (iii) power required to overcome friction in the idlers and to operate trippers, and for inclined belts, (iv) power necessary to elevate the material.

Lack of sufficient knowledge of the various friction factors together with the large number of variables affecting both the capacity and the power input requirement make the design of belt conveyors a very inexact science. The design of conveyor belts, therefore, is based largely on previous experience and empirical techniques developed by belt manufacturers (ISO/DIS 5048).

Bucket elevators

In applications where bulk solid materials have to be elevated in an area where sufficient floor space is not available for inclined belts or screw conveyors, then a bucket elevator may provide a simple alternative solution.

Essentially, the bucket elevator consists of an endless chain of metal or plastic buckets mounted on a belt or a single or double chain: the buckets are arranged to load at one end and discharge at a higher level (Figures 7.9, 7.10). Buckets may be discharged centrifugally, in which case the particles are thrown from each bucket to a discharge chute (Figure 7.11). The centrifugal action of this type of elevator requires relatively high operating

164   Mechanical conveyers

speeds of about 1.5 m/s. Such speeds can only be provided by a belt driven arrangement. Centrifugal bucket elevators are characterized by their high capacity but suffer from considerable problems of wear and tear. Alternatively, the contents of the buckets may be discharged continuously by gravity; buckets are mounted close together so that each bucket discharges on the back of the preceding one (Figure 7.11). Relatively low belt speeds of about 0.5 m/s are used with this type of elevator, which eliminates many of the problems associated with the centrifugal type.

Power requirement and discharge capacity and smooth operation of bucket elevators depend largely on speed of travel, bucket shape, size and spacing, method of feeding and discharge and the physical properties of the bulk solid material. No universally accepted method exists for the selection and design of bucket elevators for any specific duty. Selection is usually based on past experience and detailed specification of power input, belt speed and discharge capacity are obtained from manufacturers' charts.

An alternative design to the conventional bucket elevator is the air-supported belt elevator which is an extension of the air-supported belt conveyor described previously in this chapter. Essentially, in the air-supported belt elevator, the load is sandwiched between two pressurized belts that carry the material horizontally, vertically or at any inclined angle (Figure 7.12). A major advantage of this system is that the same carrying belts may be used to charge and discharge the material, thus eliminating the need for alternative conveyors for such duties. Furthermore, no separate transfer points are needed with this type of conveyor, as the horizontal belt may be extended to act as both a loading and discharging conveyor.

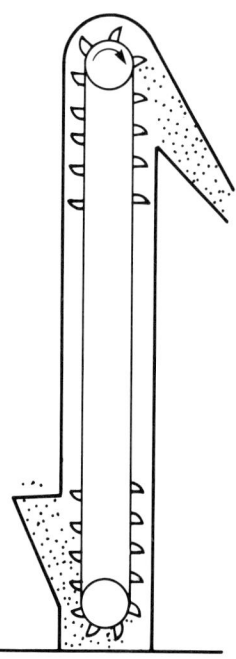

Figure 7.9 Bucket elevator.

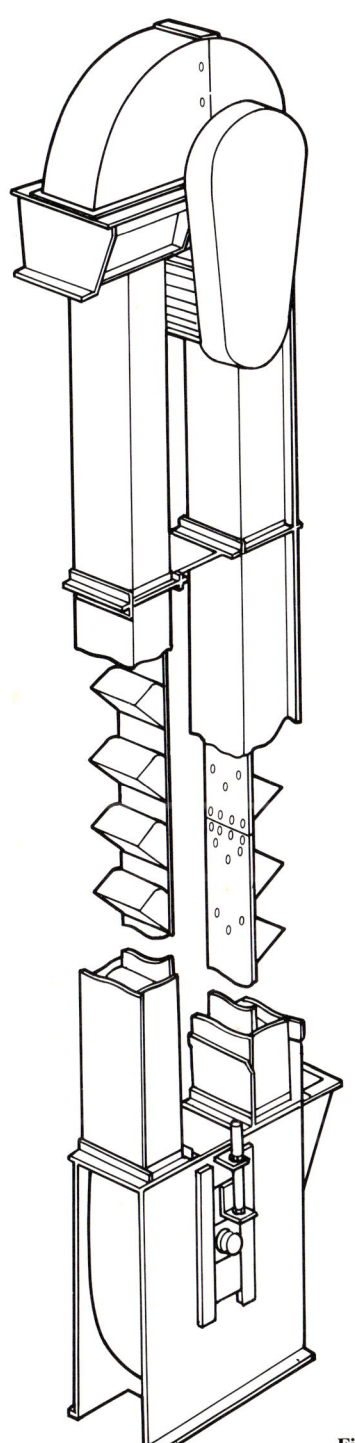

Figure 7.10 Bucket elevator.

166 Mechanical conveyers

The air-supported elevator has been used successfully to convey friable solids at capacities of up to 750 t/hr and future units have been designed to increase this to 2000 t/hr (Read, 1985).

En masse conveyors

With abrasive, friable materials, it is undesirable to use a rotating blade such as a screw to convey or elevate the solids. In such cases, the en masse or continuous flow conveyor may be employed. The en masse conveyor transports the material as a single mass inside a casing; transportation may be horizontal, inclined or vertical (Figure 7.13).

The en masse conveyor is extremely flexible in terms of loading and discharging: the conveyor is self-loading, and multi-feed and multi-discharge points may be accommodated without danger of overloading the unit. In the process industries, this type of conveyor has been used effectively to move both wet and dry chemicals and foodstuffs.

For the same load, the en masse conveyor does, however, require more power compared to other types of mechanical conveyors. The estimation

(a)

(b)

Figure 7.11 Methods of discharge from bucket elevators. (a) Centrifugal; (b) gravity.

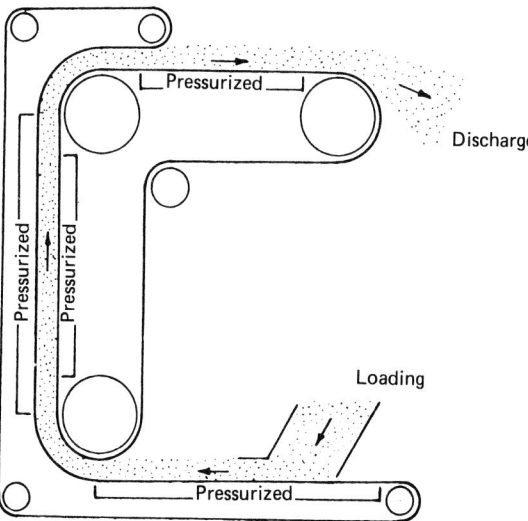

Figure 7.12 Air-supported belt elevator.

Mechanical conveyers   167

of power input requirement, operating speed, and capacity is based on empirical experience rather than on any rational method of calculation. This is largely because of the inherent variation in the properties of the bulk materials transported and also because of the very wide range of en masse conveyors commercially available.

Other conveyors and feeders

Many other conveyors and feeders are commercially available for handling bulk solids, including the pan, flight and drag conveyors (Figure 7.14) and vibratory, star and rotary plow feeders. The selection of the right type of conveyor/feeder is based largely upon previous experience and personal choice. As pointed out by Johanson (1969), such personal preferences are not always based on any objective comparison between the various options; often a single misapplication of a feeder or conveyor is enough to eliminate that particular type of unit from an entire industry.

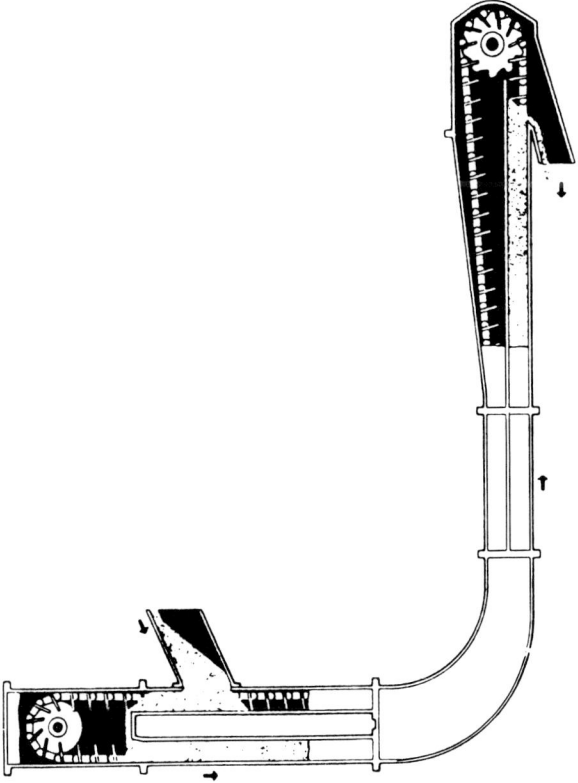

Figure 7.13 En masse conveyor/elevator.

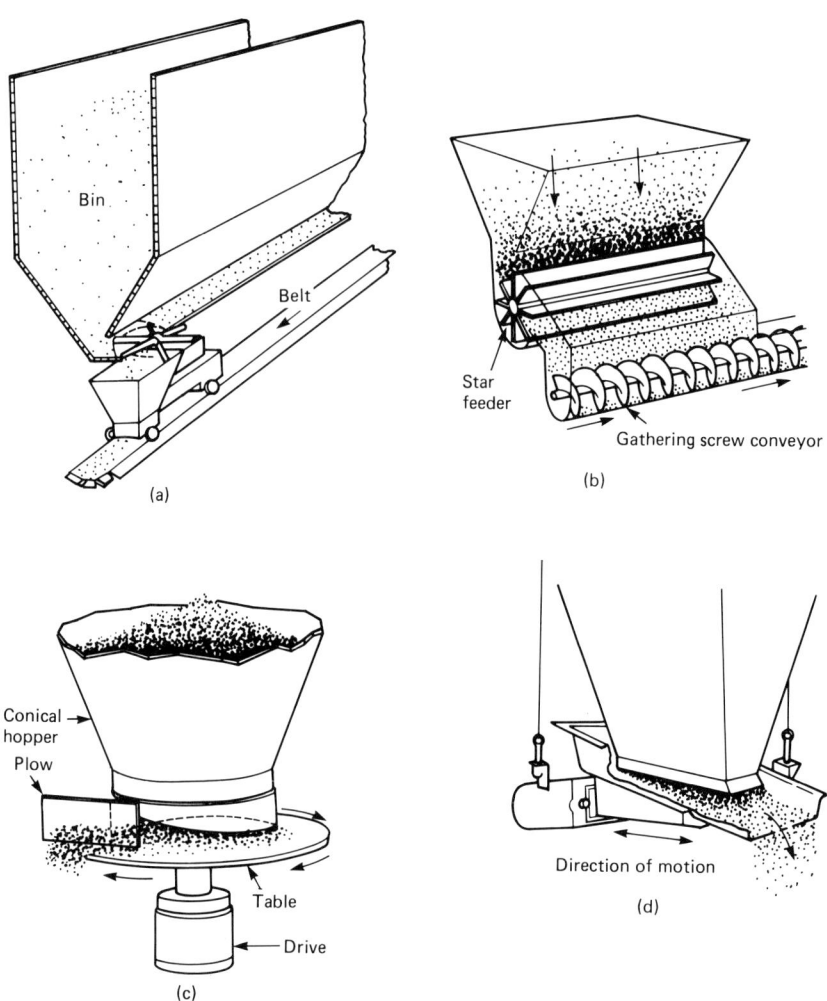

Figure 7.14 Typical feeders for bulk solids. (a) Rotary plough; (b) star feeder; (c) table feeder; (d) vibratory feeder.

References

ANON. (1985). Bulk Solids Handling, **5**, No 5, 1109
CARLETON, A. J., MILES, J. E. P. and VALENTIN, F. H. H. (1969). *J of Engng for Industry, Trans ASME*, (May), 329–334
COLIJN, H. (1985). *Mechanical Conveyors of Bulk Solids*. Elsevier Science
DOEKESEN, G. (1973). *J. Engng for Industry, Trans ASME*, **95**, Series No. 1, 93–96
ISO/DIS 5048 (1979). Draft International Standard, *Continuous Mechanical Handling Equipment*
JOHANSON, J. R. (1969). Chem. Engng. (October), 75–83
READ, G. (1985). *Bulk Solids Handling*, **5**, No. 5, 1071–1076

Chapter 8

Safety in bulk solids handling (dust explosion and health hazards)

Introduction

Apart from the usual hazards associated with the production and processing of any material in the chemical and allied industries, solids-handling plants present special dangers which require careful consideration at the design stage and constant vigilance during operation. Such dangers are many and varied, ranging from a badly designed grain silo with insufficient structural integrity that suddenly collapses during operation (Figure 2.5) to dust fires/explosions and dusts with potential health hazards. Indeed, even a non-combustible, non-toxic dust is considered to be a nuisance when dispersed in the atmosphere at concentrations greater than 10 mg/m^3 (Martin, 1975). Moreover, abrasive dust, if dispersed in air in sufficiently high concentration, could cause considerable damage to machinery.

This chapter will consider the two most serious hazards, their consequences and prevention in the general solid-materials handling and processing plants: namely dust fires/explosions and health hazards. For general information on hazards and loss prevention in the process industries the reader is referred to the relevant texts by Lee (1980a,b).

Dust fires and explosions

It is generally recognized that most combustible particulate materials can cause an explosion provided that:

(i) a sufficient amount of fine particles are dispersed in air to form a dense dust cloud;
(ii) a source of ignition is present to initiate the exothermic chemical reaction that will eventually develop into a violent explosion.

When the right conditions prevail to support such a rapid combustion reaction, large amounts of heat and hot gases are liberated. The gases expand rapidly and the pressure rises and propagates with speeds of up to 300 m/s; the sudden liberation of heat, gases and the associated pressure rise is called a dust explosion.

Ravenet (1983) points out that grains and flour dusts can be highly explosive under the following conditions:

(i) particles smaller than 200 μm;
(ii) particle concentrations of about 50 g/m^3;
(iii) source of energy equal to 1 millijoule;
(iv) temperatures of over 400° C.

The large number of reported dust explosions in grain storage silos indicates that the conditions listed above are met readily in many solids-handling plants (Dale, 1975; Theimer, 1973; Lee, 1980b; Field 1982; Lunn 1985; Schofield, 1985).

Dust generation

For a dust cloud to explode violently, the dust concentration has to fall within a critical limit. Theimer (1973) points out that below the lower explosive limit heat generated by the combustion of one particle is consumed by the surrounding air before it has a chance to reach neighbouring particles; this inhibits the rate of the combustion process, and hence the development of an explosion. Above the upper explosive limit, explosion is prevented due to incomplete combustion of particles. For grain and flour dusts, experimental data suggest that the lower explosive limit may be assumed at 20 g/m^3, but could vary between 10 and 70 g/m^3 for other dusty materials (Palmer, 1973; Schofield, 1985). The upper critical limit for flour dust is about 1000 g/m^3 and for other grain dusts is about 2000 g/m^3.

Explosive limits are affected by many variables, including particle shape, size and size distribution, moisture content, and the relative humidity of the air. Moreover, while it is relatively easy to obtain experimental values of the lower explosive limits (Palmer, 1973), the definition and determination of the upper explosive limit is far from clear (Napier, 1975).

Dust concentrations well above the lower explosive limits are generated both incidentally and intentionally in many materials-handling plants; some typical examples include:

(i) Storage silos, mixers, conveyors and settling chambers

Dust clouds may form as particles fall freely from one level to another. The air induced in the wake associated with a free-falling particle is displaced when the particle is brought to rest suddenly at the end of its travel. The resulting air movement in the wake is sufficient to dislodge dust particles from the parent particle. Once separated, the fine, dusty particles are suspended immediately in the displaced air.

The concentration of the dust cloud formed by air displacement depends upon many factors including the following:

(a) rate of material flow;
(b) height of free fall;
(c) particle size and size distribution;
(d) the free volume available, e.g. the enclosure around a conveyor belt or the free volume in a grain silo;
(e) the ease with which air can enter and leave the equipment.

(ii) Crushers, grinders, classifiers, cyclones, driers, bagging and packaging equipment, pneumatic, mechanical and manual conveyors

The exact concentration of dust produced in any specific operation is difficult to anticipate, but could vary from about 2–10 g/m^3 at a conveyor transport point to over 1000 g/m^3 for processes such as grinding. Evidently the concentration of dust clouds generated during most solids handling operations fall well within the explosive limits, and in practice, few combustible materials fail to support a dust explosion.

The range of industrial materials that can support an explosion is also wide and includes chemicals and pharmaceuticals, dyestuffs, coal, many plastics, wood, many solid foodstuffs, and agricultural and metal powders. Some metal powders do not even require the presence of oxygen to support an explosion; dispersed in either nitrogen or carbon dioxide, these dusts can produce an exothermic reaction that could result in a violent explosion (Napier, 1975).

The size of the particles determines to a very large extent the severity of an explosion. Fine powders with a mean particle size less than about 30 μm present the greatest danger; the combustion reaction is aided enormously by the large surface area provided by fine particles. Moreover, the finer the particles, the longer they stay airborne and thus the more likely it is that the dust cloud will be ignited; the amount of energy required to ignite a dust cloud also decreases with reduction in particle size. Consequently, the finer the particles, the more rapid will be the combustion process and the more violent is the resulting explosion (Cross and Farrer, 1982; Nagy and Verakis, 1983).

Sources of ignition

For a dust explosion to occur, a source of ignition must be present. The source of ignition could be any of the following:

(a) Friction heating

This could happen in almost any type of mechanical transport system such as bucket elevators, and belt and screw conveyors.

(b) Welding and cutting tools

(c) Hot surfaces

These include such plant items as radiators, steam heating pipes, and electrical appliances such as lamps.

(d) Static electricity

Powders of low electrical conductivity are particularly liable to generate static electricity. It is most likely to happen in solid materials handling equipment such as belt conveyors and dilute-phase pneumatic transport.

(e) Direct flame

For example buring cigarettes, matches and lights.

(f) Sparks

These can be due to friction between metal parts, or electrical faults, e.g. the operating of a switch in poor condition, or the blowing of a fuse.

(g) Self-heating

A pile of hot, fine powder, when unattended for a period of time, could generate enough heat to ignite a dust cloud. The self-heating phenomenon is aggravated when the pile is in an area with good air circulation. Other conditions such as dampness of the powder could initiate the self-heating process (Theimer, 1973).

Many other sources of ignition have been identified, including magnets, lightning, and the presence of foreign material. Theimer (1973) points out that even paint such as aluminium paint, containing a high percentage of a metal powder, could become a source of ignition; when a surface coated by such a paint is struck sharply by an object, an intense flash can result which can quite easily ignite a dust cloud. In this connection, it is perhaps important to point out that the above list of sources of ignition is by no means an exhaustive one and that, in fact, a substantial number of all reported primary dust explosions are caused by unknown sources; of those that are known most are caused by welding, cutting and friction (Schofield, 1985).

Explosion pressure characteristics

When a sufficient quantity of a dusty combustible material is ignited in a favourable atmosphere, large quantities of gases and heat are liberated during the exothermic runaway reaction. As a direct consequence of this, there is a rapid rise in pressure as depicted schematically in Figure 8.1. For an explosion to develop, the maximum pressure as well as the maximum rate of pressure rise are important parameters and have to exceed critical limits; for a slow pressure rise the explosion can be released safely through adequate ventilation. For a rapid pressure rise, the pressure wave generated is high enough in intensity to prevent the explosion from escaping through the normal ventilation system; the effect is often devastating, since maximum pressures of up to 1000 kN/m^2 may be generated under some conditions (a normal building cannot stand pressures much above 7 kN/m^2).

The usual chain of events leading to a major explosion is that the source of ignition will cause an initial explosion of the dust suspension. This is the so-called primary explosion and is not in itself very dangerous, except that it could generate and ignite further dust clouds by disturbing settled and usually seasoned dusts which normally accumulate all over the plants on floors, roofs, ducts, girders and machinery. The effect is often a series of secondary explosions, developing and spreading rapidly from one area to

Safety in bulk solids handling 173

another until the whole plant is affected often causing the complete destruction of the building. In many cases, the release of large amounts of heat and burning dust from the explosion also causes fires to start after the explosion. In this respect, it is worth noting that the location of a substantial proportion of all reported dust explosions between 1958 and 1975 in the USA could not be identified, presumably because of the extent of the damage; of those that could be identified, most were in bucket elevators and storage silos (Schofield, 1985).

Explosion characteristics

From what has been said so far, it may be concluded that the following parameters are important in assessing the potential explosibility of a dust cloud:

(a) lower and upper concentrations of explosive dust;
(b) minimum ignition energy;
(c) minimum ignition temperature;
(d) maximum permissible oxygen concentration to prevent ignition.

Standard tests may be carried out on samples of the dust material to evaluate the effect of the above parameters upon the severity of an explosion and to determine whether the material will explode under a given set of conditions. Such tests, however are highly empirical in nature and require considerable expertise if the results are to be meaningful (Palmer and Butlin, 1972; Palmer, 1973). As an example, Palmer and Butlin (1972) recommend the use of the vertical Hartmann tube for establishing the explosibility of a given dust cloud. In this apparatus, a known amount of the dusty material is dispersed in a flowing air stream at

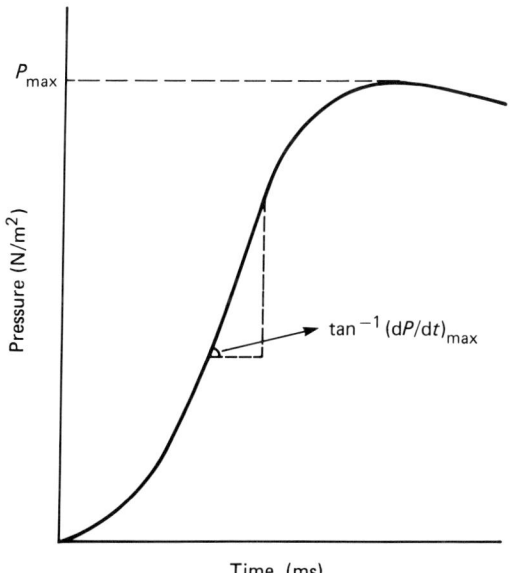

Figure 8.1 Pressure variation with time during the development of an explosion.

the bottom of the tube. The suspension is ignited by an electrical spark further up the column. The whole experiment is carried out under carefully controlled conditions and any flame propagation from the point of ignition is noted.

Based on such tests, Palmer and Butlin (1972) have classified all dusty materials into two groups as follows:

Group A. Dusts which light and propagate in the apparatus. Dusts in this group should be regarded as explosive;

Group B. Dusts which do not propagate in the test apparatus. Dusts in this group are not explosive but could be combustible.

A wide range of dusty materials have been tested and classified into Group A or Group B (Palmer and Butlin, 1972; Palmer, 1973). Any new material should be tested in the same way to establish its explosibility.

In the USA dust classification is based upon an Index of Explosibility which is defined as the product of the Explosion Severity and the Ignition Sensitivity. Explosion severity is in turn defined as the ratio of the product of maximum explosion pressure and maximum rate of pressure rise for the sample dust to that for Pittsburgh coal dust, which is used as standard. Ignition sensitivity is defined as the ratio of product of minimum ignition temperature, minimum ignition energy and minimum explosible concentration for Pittsburgh coal dust to that for the sample dust.

The index of explosibility may be used to assess the relative explosion hazard of a given dust sample using the classification shown in Table 8.1. Table 8.2 shows the effect of particle size on the index of explosibility for samples of corn starch dust (Napier, 1975).

An alternative classification of dust explosions, known as the 'St' classification, is based upon the so called 'cubic law' relationship between the maximum rate of pressure rise, $(dp/dt)_{max}$, and the volume of the

Table 8.1 Index of explosibility and classification for dust explosions (data from Palmer, 1973)

Relative explosion rating	Ignition severity	Explosion severity	Index of explosibility
Weak	0.2	0.5	0.1
Moderate	0.2–1.0	0.5–1.0	0.1–1.0
Strong	1.0–5.0	1.0–2.0	1.0–10
Severe	5.0	2.0	10

Table 8.2 Variation of Index of explosibility with particle size for Cornstarch (Napier, 1975)

Mean particle size (μm)	Ignition sensitivity	Explosion severity	Index of explosibility
130	0.4	0.2	0.1
90	0.9	0.9	0.8
48	3.8	9.8	37
22	4.3	23.2	100

vessel, V, (Field, 1982; Schofield, 1985). This law may be expressed by the following equation:

$$\left(\frac{dp}{dt}\right)_{max} = KV^{-1/3} \tag{8.1}$$

where K is a constant characterizing the explosion (Table 8.3).

Explosion prevention

In principle, it is easy to prevent an explosion; all that is needed is to avoid dust accumulations above the minimum explosive concentration and eliminate all sources of ignition. In practice it has proved very difficult to achieve these, as demonstrated by the rising number of reported dust explosions (Dale, 1975; Lee, 1980a,b). The cost of such explosions is considerable, both financially and in terms of human life. For example, a series of dust explosions occurred in a set of grain silos at Westwego near New Orleans, Louisiana, in 1977, killing 35 people (Figure 8.2). Lee (1980a,b) points out that this disaster was on the same scale as that of Flixborough.

Figure 8.2 Dust explosion of grain silos at Westwego, USA, 1977.

Table 8.3 St Classification of dust explosions (data from Schofield, 1985)

Relative explosion rating	K (see Equation 8.1) kN m/s	St Classification
Severe	> 300 000	3
Strong	200 000 < K < 300 000	2
Weak	0 < K < 200 000	1
No explosion	0	0

It is therefore reasonable to assume that dusts will always exist above the minimum explosive limits, and that sources of ignition are always present. However, a good deal can be done to cut down the amount of dust deposited and reduce the possibility of an ignition. Furthermore, if after all foreseeable measures have been taken to stop the formation and ignition of a dust cloud, a fire or primary explosion occur, preventive actions should be taken to suppress the progress of the spreading flames and the development of secondary explosions.

Inhibiting dust formation and accumulation

Since it is difficult to prevent the generation of dust without seriously affecting the performance of the plant, the best solution is to install permanent exhaust and venting systems for dust, provided it is not toxic. The venting area required increases with increase in the severity of the explosion. Palmer and Butlin (1972) have provided an empirical guide for selecting the required vent area (Table 8.4).

The information given in Table 8.4 is based upon pressure measurements in the Hartmann apparatus, and appears to overestimate the required vent area. Alternative methods have been proposed to evaluate the required vent area which are claimed to be more realistic (Field, 1982).

One such approach used in USA for the estimation of the minimum vent area is based on experimental data from a spherical apparatus with a minimum volume of 20 litres. The results from this device can be used to determine the explosion characteristics of the dust, that is its 'St' classification: this information may then be applied directly for scale up purposes using the cube-root law (Equation 8.1). Thus, knowing the St class of a given dust, or the K value in Equation 8.1 (Table 8.3), and the exact dimensions of the spherical apparatus, that is its volume and area, it is possible to calculate the required vent area for any other vessel volume within the plant handling the same dust.

The detailed calculation of the required vent area by this method must also take into account other factors such as the strength of the vessel, the pressure at which the explosion relief cover is fully open and the sources of ignition: the actual calculations have been made a great deal easier by the use of nomographs provided by NFPA 68 (Anon., 1978) and VDI 3673 (Anon., 1979). The charts which cover the range of operating conditions of interest in most practical situations may be used to size vent areas for vessels of up to 1000 m^3, provided the ratio of length to scale of the container is below or equal to 5.

Table 8.4 Recommended vent ratios for containers up to 30 m^3 (data from Palmer and Butlin, 1972)

Maximum rate of pressure rise (kN/m^2)	Vent ratio (vent area/plant volume) (m^{-1})
< 35 000	1/7
35 000–70 000	1/5
> 70 000	1/3

Dust should be collected and removed from any dust-forming equipment as early as possible. Dust collectors, e.g. cyclones and bag filters, should be used, but with care; the dust collected in such units should be discharged continuously and, if the stored material is explosive, there should be no possibility of any flame path between the process equipment and the collector hopper (Green, 1975).

Dust clouds are generated by air displacement whenever particles fall freely into storage silos, mixers and screens. When possible, the inlet to such units should be securely enclosed and dust control equipment installed. Furthermore, the available free volume in such equipment should be kept to a minimum.

The feed equipment, say, from a storage silo to a conveyor belt should be designed so that particle free fall is minimized.

Building and equipment layout should be considered with a view to minimizing dust collection and accumulation. For example, every effort should be made to separate dusty processes from non-dusty operations in order to reduce the size of the contaminated area to a minimum. All areas between equipment and walls should be easily accessible for cleaning. Doors, frames, floors, girders, beams and ledges should not be permitted to act as areas where dust might be trapped.

Prevention of ignition

Under no circumstances should a direct flame such as a burning cigarette or matches be allowed in areas of the plant where explosive dust clouds might be present. All other sources of ignition such as welding and cutting tools should only be allowed in dusty areas when it is authorized by a safety officer; the plant should be thoroughly cleaned and vented before any repair work is carried out.

In any dust-laden area there should be no source of indirect heat such as hot surfaces, e.g. radiators and electric lamps. All faults such as can occur with the bearings of motors, bucket elevators and electrical appliances should be reported to the safety officer as soon as they are detected; such equipment should be checked regularly and defective parts replaced immediately.

Hot, fine powders from any part of the operation, e.g. driers, should be cooled before storage in order to minimize chances of ignition due to spontaneous heating. A temperature-sensing device may be placed at a suitable location in the storage vessel in order to detect any increase in temperature due to the self-heating process; appropriate action should be taken as soon as this is observed.

To mimimize danger of ignition due to static electricity, major items of equipment should be grounded, preferably to a standard source; this earthing should be inspected at regular intervals.

Finally, each plant should be considered carefully in order to identify all possible sources of ignition; appropriate instructions should be made to all operators to minimize the chances of an explosion. Operator-training is essential in order to ensure that the necessary precautions are taken in case of emergency. For example, no metal objects or electrical lamps should be lowered into any solids storage bin (Theimer, 1973); for inspection purposes, battery-operated handlamps only should be used.

Explosion suppresion

In practice, the time interval between ignition and maximum pressure rise is relatively small, ranging between tens to hundreds of milliseconds. However, if a developing explosion is detected before the maximum pressure is reached, the explosion can, under certain conditions, be arrested by using an explosion suppresion system.

Essentially, an explosion suppression system consists of an automatic sensing instrument that detects changes in, say, pressure within the plant. Once a pressure-wave is detected, a suitable chemical extinguishant is injected automatically and at a high speed to suppress the explosion flame and thus stop the pressure rising to its maximum explosive limit (Figure 8.3). The suppressant, which is an inert substance (gas, liquid or powder), is injected from containers under pressures of up to 300 p.s.i.g. The containers are normally mounted on fairly rigid walls immediately outside the protected zone; the number, capacity and exact location of the suppressors should be selected with care to ensure that the affected area is filled with the extinguishant in the shortest possible time.

For the suppression action to be effective, the time delay between detection and injection of the suppressant has to be greatly in excess of the velocity of the spreading flame (Table 8.5). To achieve the necessary response time, the injection process is initiated by using a suitable explosive which is electrically detonated; the delay time in such a system is less than 1 millisecond from the moment that the detector contacts are closed.

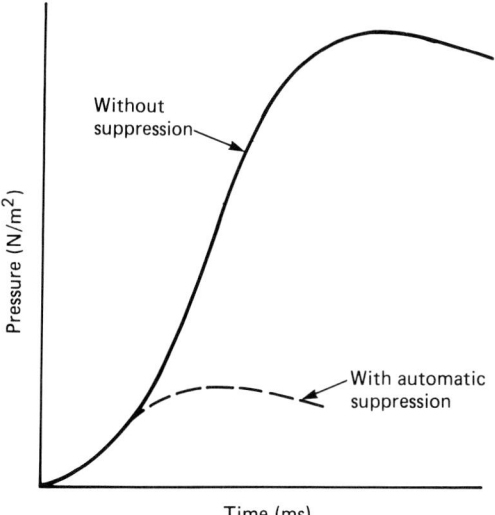

Figure 8.3 Effect of an automatic suppression system upon the variation of pressure with time during the development of a dust explosion.

Lee (1980a,b) points out that explosion suppression is most effective for container volumes up to 115 m³, and for flames for which the maximum pressure rise is reached in not less than 40 ms. Moore (1986) claims that by careful selection of suppressors, volumes of up to 1000 m³ may be treated effectively by this method. Moore also confirms that the necessary number of high-discharge rate suppressors is related to the volume of the container and may be estimated using the cubic law relationship:

$$n = n_0(V)^{2/3} \qquad (8.2)$$

where n is the number of suppressors required to suppress an explosion in volume V, and n_0 is the fictitious number of the same suppressors needed for a similar explosion in a volume of 1 m³.

In conclusion, it is essential that the process and design engineers handling dusty powders should seek expert advice, particularly when the materials are potentially explosive. In the UK, such advice may be obtained from a number of sources, including the Fire Research Station, Borehamwood, and the Hazards and Process Group, ICI, Organics Division, Blackley.

Health hazards

Finely divided particles dispersed in air can enter the human body by inhalation, swallowing or by direct contact, e.g. with skin, eyes or mouth. Dust inhalation is, however, the most common mode of entry and is by far the most damaging to the recipient. This is readily appreciated when it is realized that while on average we eat only three times a day, we breath over 7000 times during an eight-hour working day and inhale more than a million particles of a size visible in the optical microscope.

What happens to a dust particle when it is inhaled depends to a very large extent upon its size and density. Particles with diameter in the range 15 to 25 μm are filtered out in the nasal passage or are captured by direct impact against the moist wall of the respiratory tubes. Once caught, these particles are moved toward the exits of the nose and mouth by the action of

Table 8.5 Linear burning velocities in dust trains (data Palmer, 1957)

Material	Particle size (μm)	Linear burning velocity (m/s)
Coal	< 104	20
Cocoa	< 40	27
Starch	< 40	40
Sawdust (dry)	< 250	670
Sawdust (dry)	250–1200	280
Beech sawdust	48 (mean)	15.1
	27 (")	15.4
	19 (")	16.1
	14 (")	16.8
	12 (")	17.0

of cilia cells; these are fine, hairlike cells covering the respiratory passages and capable of independent movement.

Particles with diameter in the range 0.1 to 1.0 μm are usually exhaled from the body during the normal expiration.

Particles smaller than 0.1 μm present the greatest dangers, as they can reach deep into the lungs and could eventually enter the blood stream. The hazardous effects on the human body depend upon factors such as the type and concentration of the dust and the period of exposure. For example, silicosis, which is fibrosis of the lungs, is caused by inhalation of quartz (SiO_2) particles. Some of the symptoms of this disease, like shortness of breath, do not appear until several years after exposure. For the first stage of the disease to develop, however, a period of exposure of about 8 months is needed (Drinker and Hatch, 1954). If the period of exposure or the quartz-dust concentration is very high, the disease becomes complicated by tuberculosis of the lungs.

Another well-know form of fibrosis of the lungs is asbestosis caused by inhalation of asbestos dust; in some cases exposure can also lead to cancer of the lungs. It is difficult to decide with any certainty the length of exposure and the required dust concentrations to cause the disease; it is generally believed that several years of exposure in high dust concentration is needed for the disease to develop, but in some cases relatively short periods of exposure have been shown to cause asbestosis (Lee, 1980).

It is also possible for the finer dust particles to find their way into the blood stream after they reach the lungs. In this way these particles can reach all parts of the body quickly and effectively. Typical examples of poisoning of the blood by toxic dusts include lead, cadmium oxide, manganese, chromates, arsenic, insecticides, and radioactive dusts.

The assessment of the effect of toxic materials, including dusts, is essentially a medical matter (Lee, 1980a,b). It is therefore important for the process and design engineers involved with handling such materials to seek specialist advice on questions such as long- and short-term exposure and the maximum permissible hazardous dust concentrations for safe operation under a given set of conditions; in practice, the aim must be to meet these requirements and, if possible, to go further and minimize dust concentrations as far as practicable.

In the UK, the following sources can provide some useful information on the hazards of dusts:

1. Fire Research Station, Borehamwood.
2. Department of Employment, London (HM Factory Inspectorate)
3. Department of the Environment, London (HM Stationery Office)
4. Health and Safety Executive, London (HM Factory Inspectorate)
5. The Institution of Chemical Engineers, Rugby.

Dust control equipment

In most industrial bulk-solid handling plants, it is difficult to suppress dust formation at source without seriously affecting the performance of the plant. The best solution to deal with dust problems is therefore to reduce and extract the dust from the atmosphere in the affected zones. It is unlikely that the dust can be eliminated completely and the aim should be to reduce the concentration of the dust to below acceptable levels. This usually involves installation of gas-cleaning and dust-collection equipment at all dust emission points within the production plant: for maximum protection, the dust-control equipment can be interlocked with the machinery so that it is impossible to start production until the exhaust facility is energized.

There are many types of dust-control equipment, from the simple dust hood (Figure 8.4) to fabric and fibrous filters, reverse jet filters (Figures 8.5, 8.6), wet and dry precipitators, cyclones (Figure 8.7) and scrubbers (Figure 8.8). The choice depends largely upon the nature of the dust to be collected, e.g. smallest particle size and density to be captured, gas temperature, the volume of the contaminated air to be treated, collection efficiency and the economics of the operation. For a given type of dust-control equipment, collection efficiency drops sharply as particle size decreases: the plot of collection efficiency versus particle size for a given collector is known as the grade efficiency for that collector and is used as a guide to selection of dust control equipment (Figure 8.9 and Table 8.6).

Other factors that need to be considered before the final selection of the type of dust-control system include whether or not the dust is sticky, fluffy, erosive, toxic or combustible, that is prone to explosion (see Table 8.7). For example, sticky dusts could easily blind most positive types of filters and therefore wet collectors are generally more suitable for such materials (Nonhebel, 1972).

The proper design and selection of fail-safe and efficient dust-control equipment requires a detailed study of the nature of the dust involved and the individual items of equipment (Chambers, 1985, 1986a) as well as the overall production, processing and handling methods involved (Chambers 1986b; Muir, 1985). Very often a piece of dust-control equipment such as a filter is designed for a single or closely related group of dusts and for a specific point of emission.

Because of the variety in choice and the importance of economic operation of dust-control equipment, the selection and detailed design of such machineries should always be carried out in close consultation with an experienced dust-control engineer if failure and, in some cases, disaster is

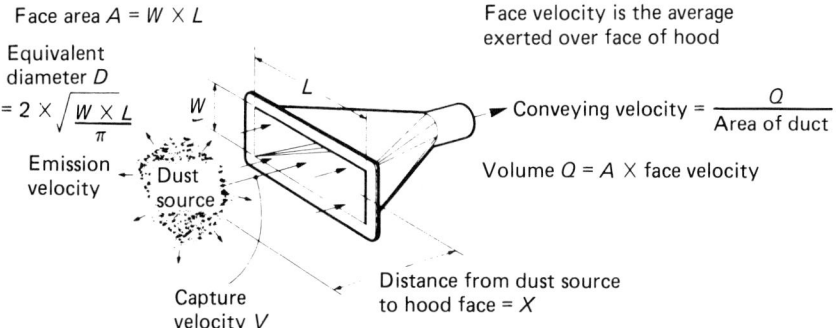

Figure 8.4 Principle of operation and design of dust collection hoods.

182 Safety in bulk solids handling

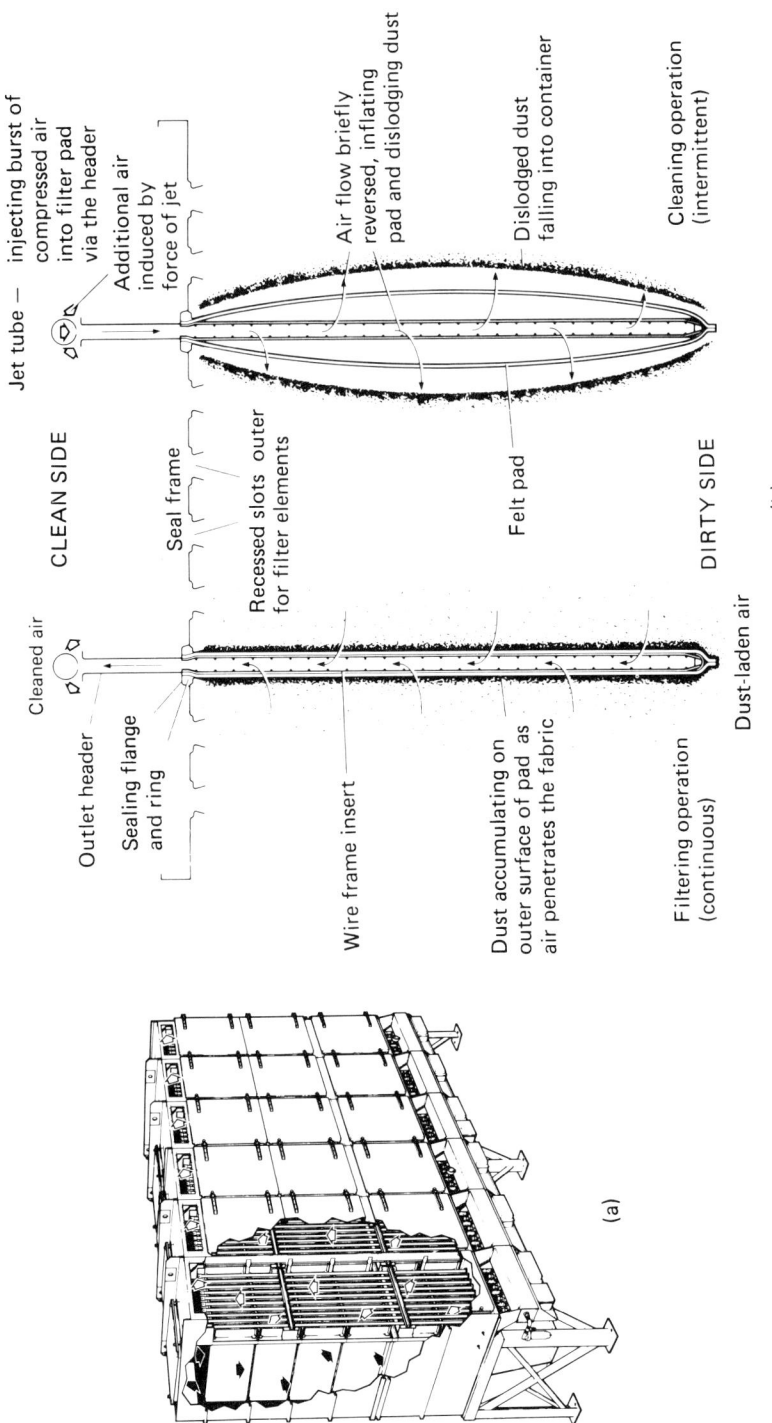

Figure 8.5 Typical continuous-rated reverse-jet fabric filter. (a) Layout of frames; (b) principle of operation showing continuous filtering and intermittent cleaning.

Table 8.6 Efficiency of dust collectors (data from Muir, 1985)

Type of Collector	Efficiency			
	at 10 μm	at 5 μm	at 2 μm	at 1 μm
Inertial collector	30	16	7	3
Medium-efficiency cyclone	45	27	14	8
High-efficiency cyclone	87	73	46	27
Low resistance cellular cyclone	62	42	21	13
Tubular cyclone	98	89	77	40
Irrigate cyclone	97	87	60	42
Self-induced spray deduster	98	93	75	40
Spray tower	97	94	87	55
Wet impingement scrubber	> 99	97	92	80
Disintegrator	99	98	95	91
Venturi scrubber-medium energy	> 99.9	99.6	99	97
Venturi scrubber-high energy	> 99.9	99.9	99.5	98.5
Electrostatic precipitator	> 99.5	> 99.5	> 99.5	> 99.5
Irrigated electrostatic precipitator	> 99.5	> 99.5	> 99.5	> 99.5
Shaker-type fabric filter	> 99.9	99.6	99.6	99
Pulse-jet fabric filter	> 99.9	99.6	99.6	99.6

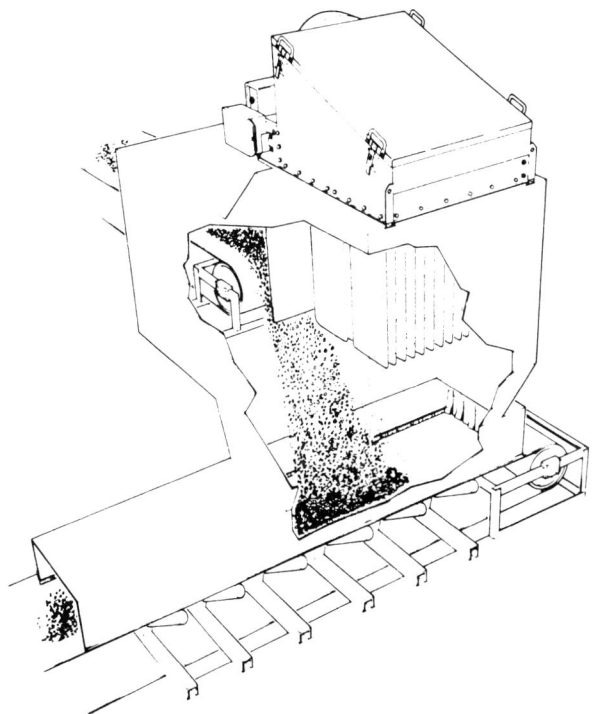

Figure 8.6 Dust filter incorporated in conveyor transfer point.

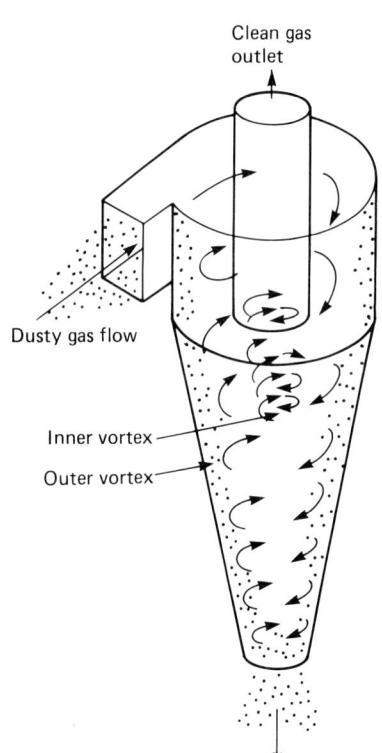
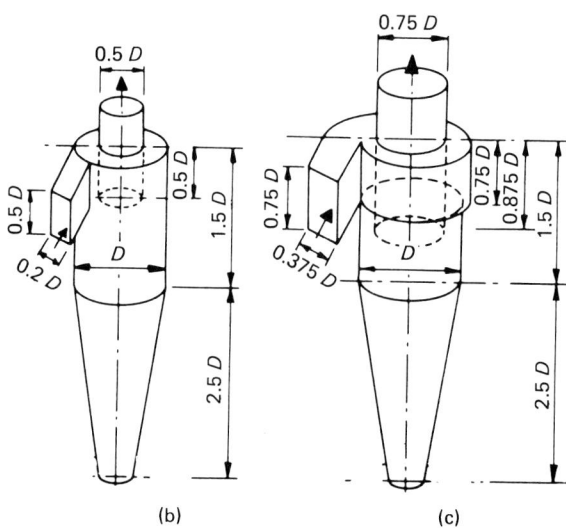

Figure 8.7 Standard cyclone operation and design. (a) Principle of operation; (b) high-efficiency design with medium throughput: flow rate 1.5 $D^2 m^3/s$; (c) medium-efficiency design with high throughput: flow rate 4.5 $D^2 m^3/s$. Typical gas flow rate for (b) and (c): 15 m^3/s. D = diameter (m).

Safety in bulk solids handling 185

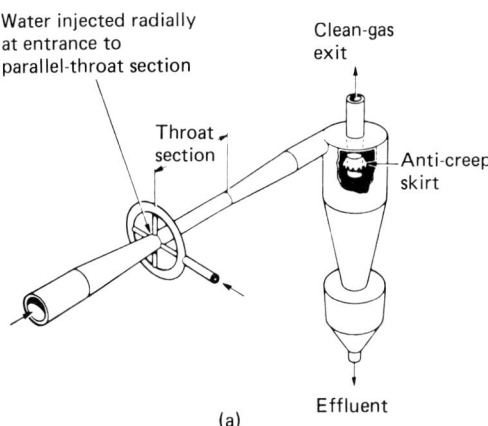

(a)

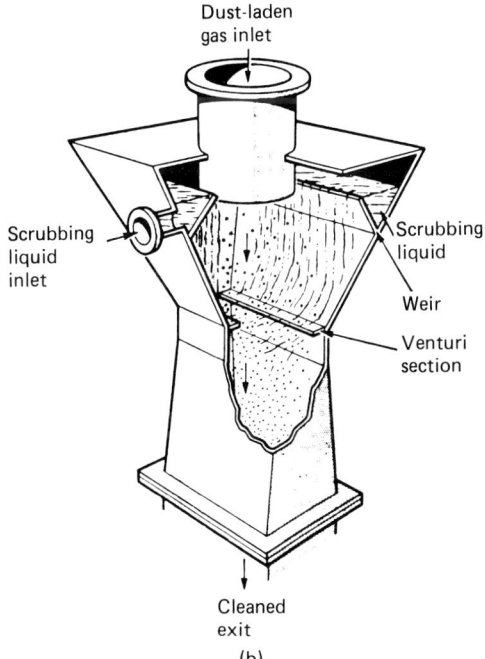

(b)

Figure 8.8 Wet dust collectors (venturi scrubbers). (a) Basic arrangement; (b) arrangement for recirculating scrubbing liquid.

186 Safety in bulk solids handling

to be avoided: this is particularly important when the dust to be captured is potentially explosive and/or toxic (Figures 8.2 and 8.10).

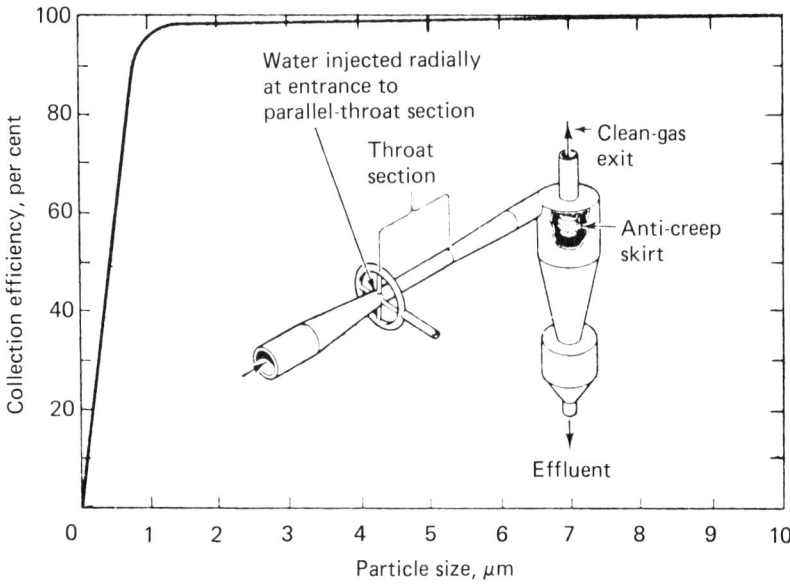

Figure 8.9 Grade-efficiency curve for venturi scrubber (Nonhebel, 1972).

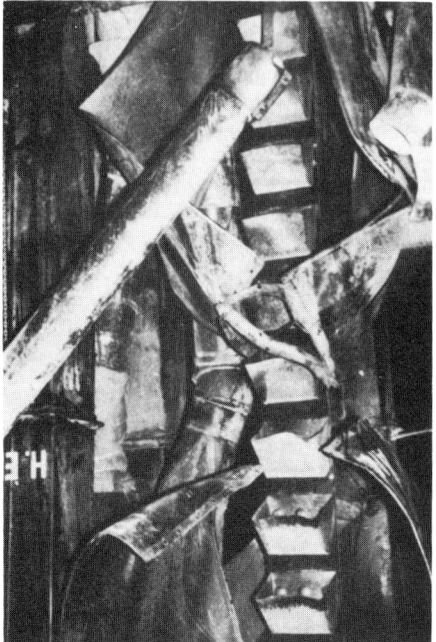

Figure 8.10 Bucket elevator after dust explosion of grain silos at Albern, Austria.

Table 8.7 Guide to selection of equipment for dust control (Muir, 1985)

	Dust properties							Gas conditions					Other factors		
	High inlet burden	Ero-sive	Sticky	Light fluffy	Diffi-cult to wet	Pyro-phoric	Resist-ivity problem	Const-ant press drop	Vary-ing flow	Explos-ive Combust-ible (1)	Corro-sive	Suitable for high pressure	Minim-um ancil-llary equipment	On-line regen-eration
Cyclones	✓	✓	beware	beware	✓	care	✓	✓	care	care	care	✓	✓	✓
Wet washers Low energy	✓	✓	✓	✓	care	✓	✓	✓	care	✓	care	✓	care	✓
Wet washers High energy	✓	care	✓	✓	care	✓	✓	✓	care	✓	care	✓	care	✓
Dry electrost-atic precipit-ators	care	✓	care	care	✓	care	beware	✓	care	beware	care	care	care	
Wet electrostatic precipitators	care	✓	✓	✓	care	✓	care	✓	care	beware	care	care	care	✓
Aggregate filters	care	✓	beware	✓	✓	✓	✓	care	✓	care	care	✓	care	care
Fabric filters	✓	care	beware	✓	✓	beware	✓	care	✓	care	care	care	care	care
Fibrous filters	beware	✓	care	✓	✓	care	✓	care	care	care	care	✓	✓	beware

Key
✓ Indicates that the type of plant can generally cope with the process requirement, if well designed.
Care Indicates that special attention is required in both plant design and operation to prevent problems.
Beware Indicates that this process condition could lead to severe operational difficulties and for this reason other alternatives which avoid the problem are normally sought.
(1) For all gas cleaning problems associated with explosive or combustible materials, competent advice should be sought.

References

ANON. (1978). *Guide for Explosion Venting*. NFPA No. 68, National Fire Protection Association, Boston
ANON. (1979). *Pressure Release for Dust Explosions*. VDI Richlinien 3673, VDI-Verlag GmbH, Dusseldorf
CHAMBERS, D. (1985). Safecast 85 Supplement
CHAMBERS, D. (1986a). *Manufacturing Chemist*, October issue
CHAMBERS, D. (1986b). *A Practical Guide to Dust Control*. DCE Group literature
CROSS, J. and FARRER, D. (1982). *Dust Explosions*. Plenum Press
DALE, S. G. (1975). Powtech 75. *Powder Technol.* Series No. 6, 59–61
DRINKER, P. and HATCH, T. (1954). *Industrial Dust*, 2nd ed. McGraw-Hill
FIELD, P. (1982). Dust Explosions, *Handbook of Powder Technology*. Elsevier Scientific
GREEN, K. D. (1975). *Powder Technol.* Series No. 6, 76–83
LEE, F. P. (1980a). *Loss Prevention in the Process Industries*, Vol. 1. Butterworth, London
LEE, F. P. (1980b). *Loss Prevention in the Process Industries*, Vol. 2. Butterworth, London
LUNN, G. A. (1984). Venting Gas and Dust Explosions – A Review, I. Chem. E.
MARTIN, R. (1975). *Powder Technol.* Series No. 6, 81–84
MOORE, P. (1986). *The Chemical Engineer*, No. 430, pp. 43–47
MUIR, D. M. (ed.) (1985). *A User Guide to Dust and Fume Control*, I. Chem. E.
NAGY, J. and VERAKIS, H. C. (1983). *Development and Control of Dust Explosions*. Marcel Dekker, Inc. New York
NAPIER, D. H. (1975). *Powder Technol.*, Series No. 6., 51–56
NONHEBEL, G. (1972). *Gas Purification Processes*, 2nd Edn., Butterworth, London
PALMER, K. N. (1973). *Dust Explosion and Fires*, Chapman and Hall
PALMER, K. N. and BUTLIN, R. N. (1972). *Powder Technol.*, **6**, 149–157
RAVENET, J. (1983). *Bulk Solid Handling*, Vol. 3, No. 1, 127–141
SCHOFIELD, C. (1985). *IChemE Guide to Dust Explosion Protection*, Part I
THEIMER, O. F. (1973). *Powder Technol.*, **8**, 137–147

Author index

Abramovitsch, G.R., 26
Ahumada, J.J., 46
Al-Din, N., 80, 85
Allen, T., 3, 8
Aoki, R., 38, 39
Arnold, P.C., 34, 40, 43, 56, 59, 60
Aude, T.C., 145, 146, 147

Babcock, H.A., 134, 135
Bachouchin, D.M., 112, 116
Bailey, A.G., 16
Barnsby, P.L., 46
Barth, H.W., 112
Basur, H.S., 42, 43
Beverloo, W.A., 81, 83, 84, 85, 88, 91
Birks, A.H., 63
Bishara, A.G., 31
Blachard, M.H., 38
Borg, L., 63
Bowen, R.LeB., 144
Bragg, G.M., 107
Broadhurst, P.M., 107
Brooks, D.A., 132
Brown, R.L., 81, 82, 83, 85, 87, 88, 90, 93
Butlin, R.N., 173, 174, 176

Carleton, A.J., 80, 89, 91, 92, 161
Carleton, R.P., 138, 144
Carr, L.Jr., 16
Chambers, D., 181
Chapus, E.E., 134
Cheng, D.C.H., 138, 144, 145
Chhabra, R.P., 136
Chong, Y.O., 116
Clift, R., 10, 12, 13
Colijn, H., 156
Condolios, E., 134
Cowin, S.E., 22, 28
Cross, J., 171

Dale, S.G., 170, 175
Dalla Valle, J.M., 107
Dalstad, J.I., 146
Darby, R., 142
Darton, R.C., 93
Davidson, J.F., 80, 88, 89, 93
Deming, W.E., 85, 93, 94, 95, 96
Deutsch, G.P., 31, 44
Dia Felira, R., 14
Dodge, D.W., 143
Dodweel, C.H., 132
Doeksen, G., 157
Drinker, P., 180
Duckworth, R.A., 117
Durand, R., 134

Eckhoff, R.K., 73
Edwards, M.F., 16
Enstad, G., 31, 59
Evans, A.C., 80, 92
Everts, R., 33

Faddick, R.R., 131
Farrev, D., 171
Field, P., 170, 174, 176
Foivlev, P.J., 80, 84, 98
Foscolo, P.V., 14
Foster, P.J., 80, 84, 98
Franklin, F.C., 80, 87

Galli, A.F., 110
Gau, G., 116
Gaylord, C.N., 73
Gaylord, E.H.Jr., 73
Geldart, D., 4, 12, 13, 121
Gewdson, B.J., 91
Gibilaro, L.G., 14
Glastonbury, G.R., 80, 84, 85
Glunta, J.C., 43, 44

Godfrey, A.R., 103
Govatos, G., 136, 137
Gravestock, N., 91
Green, K.D., 9, 177
Griffith, J., 40
Gunn, D.D., 80, 85

Hancock, A.W., 31
Haris, E.C., 73
Hariu, O.H., 111, 113
Harmens, A., 81, 82, 83, 84
Harnby, A., 11, 132
Harr, M.E., 28
Harris, R.W., 144
Hass, D.B., 136
Hatch, T., 180
Hayden, J.W., 135
Haze, H., 76
Hazra, S.K., 42, 43
Head, J.M., 46
Heywood, H. 4, 5
Higman, R.W., 145
Hinkle, B.L., 112, 113, 114
Horne, R.M., 31
Huggett, M.R., 103

Ikemori, K., 112

Janssen, H.A., 22, 25, 26, 31, 32, 33, 34, 35, 39, 41, 43, 44, 70
Jenike, A.W., 15, 23, 25, 28, 31, 33, 34, 39, 40, 41, 43, 51, 52, 55, 56, 57, 60, 61, 65, 70, 72, 73, 81, 83
Johanson, J.R., 25, 31, 60, 70, 157, 161, 163, 167
Johnson, L.N., 80, 87, 99
Jones, A.G., 90
Jones, P.S., 115, 116
Junior, C.C., 31

Kenchington, J.M., 144
Klinzing, G., 10, 12, 112, 117, 118
Knepper, W.A., 16
Knowlton, T.M., 112, 116, 121
Konno, H., 112
Kotchanova, I.I., 82
Kraus, M.N., 103, 105
Kuno, H., 93, 98
Kvapil, R., 44
Kwan, M.Y.M., 107

Lazarus, J.H., 132, 143, 144
Lee, F.P., 169, 170, 175, 179, 180
Leser, T., 56
Leung, W.C., 106, 107, 112, 115, 116, 118, 119

Leversen, G., 73
Lipnitskii, M.E., 26
Liptak, B.G., 147
Lord, D.L., 144
Lunn, G.A., 170

Martin, R., 169
Mason, J.S., 104
Matsumoto, S., 116
McCabe, W.L., 109
McDougall, J.R., 80, 92
McLean, A.G., 34, 40, 56, 59, 60
McLeman, M., 112, 118
Mehring, A.L., 85, 93, 94, 95, 96
Memon, M.A., 80, 84, 98
Metzner, A.B., 142, 143
Miles, J.E.P., 91, 92, 94, 96, 97
Molerus, O., 13
Molstad, M.C., 111, 112
Mooij, A., 33
Moore, P., 179
Muir, D.M., 181, 183
Murfitt, P.M., 42
Murphy, J., 84

Nagy, J., 171
Napiev, D.M., 170, 171, 174
Nedderman, R.M., 31, 80, 81, 82, 83, 84, 85, 88, 89, 90, 91, 92, 93, 99
Newitt, D.M., 133, 135, 136
Newton, R.H., 81
Nonhebel, G., 181
Nienow, A.N., 16, 46

Ooms, M., 41, 43, 99
Orr, C.J., 11, 130
Owens, P.R., 115

Palmer, K.N., 170, 173, 174, 175
Papazoglou, C.S., 93
Peters, L.K., 118
Pfeffer, R., 118
Pyle, D.L., 93

Rappen, A., 75
Ravenet, J., 22, 31, 53, 169
Read, G., 162, 163, 166
Reimbert, V.A., 25, 27, 28, 29
Reznich, W., 92
Richards, P.C., 33, 45, 72, 81, 82, 83, 85, 87, 88, 90, 93
Richardson, J.F., 112, 118, 136
Rizk, F., 115, 116
Roberts, A.W., 43, 99
Roher, J.H., 41
Rose, H.E., 9

Author Index

Rose, H.F., 80, 87, 117
Rossetti, S.J., 118
Rowe, P.N., 91
Rumpf, H., 11, 12

Safarian, S.S., 73
Saito, S., 112
Savage, S.B., 80
Savrousky, L., 132
Schmidt, J., 31, 44
Schofield, C., 170, 172, 173, 174
Schubert, H., 13, 55
Schuchart, P., 113
Schwedes, J., 30
Seville, J.P.K., 13
Shamlou, P.A., 131, 138, 141
Simons, H.P., 118, 119
Sive, A.W., 143, 144
Smith, J.C., 109
Sokolovski, V.V., 25
Soo, S.L., 118
Spinks, C.D., 82, 90
Stelson, T.E., 135
Stepanoff, A.T., 84, 102, 134
Stoess, H.A.Jr., 105
Streat, M., 132
Sugden, M.B., 41
Sundram, V., 22, 28

Tamura, T., 76
Tanaka, T., 80, 87

Theimer, O.F., 170, 172, 178
Trezek, G.J., 118
Tsunakawa, H., 38, 39
Turian, R.M., 134, 137, 145
Tuzun, V., 90, 99

Van Zanien, D.C., 33, 44
Verakis, H.C., 171

Waight, H., 73, 75
Walker, O.H., 25, 31, 35, 38, 40, 41, 55, 72
Walters, J.K., 25, 35, 40
Wen, C.Y., 110, 111, 112, 117, 119
Whetton, J.T., 107
Wiles, R.J., 106, 107, 112, 118, 119
Wilkinson, W.L., 142
Williams, J.C., 41, 67, 75, 80, 82, 88, 89, 91
Wood, J.G.M., 41

Yates, J.G., 93
Yang, W.C., 107, 111, 112, 113, 114, 116
Yousifi, Y., 116
Yu, Y.H., 110
Yuan, T., 134, 137
Yuasa, Y., 93, 98

Zandi, I., 136, 137
Zenz, F.A., 114, 115

Subject index

Agglomerate strength, 12
Aerated density, 16
Air cannon, 75
Air lance, 75
American ACI Committee, 26
Angle of internal friction, 28
Angle of wall friction, 65

Belt conveyors, 161
Belt feeders, 163
Blow tanks, 120
  air bypass pulse-phase, 120, 121
Bowen's design equations, 144
Bucket elevators, 163
Bulk density, 16, 53
  aerated density, 16
  average density, 16
  fluid density, 16
  loose density, 16
  packed density, 16
  tapped density, 16
  working bulk density, 16

Capillary forces, 9, 11
Centrifugal bucket elevators, 164
Choking, 107
Choking velocity, 114

Dense phase flow, 106, 118, 119
Dilute phase flow, 106, 108
Drag conveyors, 166
Dust, 169
  classification, 174
  control equipment, 180
  explosion, 169
    characteristics, 173
    prevention, 175
    suppression, 178
  fire, 169

Dust(*cont.*)
  generation, 170
  health hazards, 179
Dynamic discharge pressures, 31

Effective angle of internal friction, 65
Electrostatic forces, 9, 10
En-masse conveyors, 166
Expanded flow, 22
Explosion severity, 174

Flammability, 2
Flow/no flow criterion, 52
Flow factor, 63, 65
Flow function, 63, 65
Flow-promoting devices, 75
Flow regimes, 132
Flowability, 2
Fluidizability, 2
Funnel flow silos, 20, 22, 42, 51, 73
  design of, 51, 73
  pressure transmission in, 41

Geldart's classification, 12, 13
German DIN 1055, 26, 32
Gravity discharge, 80
  of coarse particles from bins, 84
  of coarse particles from silos, 86
  of fine particles, 91

Heterogeneous flow, 135
Homogeneous flow, 105, 132
Hydraulic conveying, 132
  flow regimes, 132
  pressure drop, 133
  transition velocity, 135

Ignition sensitivity, 174
Index of explosibility, 174

Subject Index

Janssen's formula, 25
Jenike shear cell, 11
Jenike's classification, 16, 55
Jenike's strain–energy theory, 34

Linear mean diameter, 8
Lower explosion limit, 170

Major consolidating stress, 15
Mass flow, 20, 22, 31
Material yield locus, 14, 70
Mean surface diameter, 8
Mechanical arching, 52
Mechanical conveyors, 157
  belt conveyors, 161
  belt feeders, 163
  bucket elevators, 164
  centrifugal bucket elevators, 164
  screw conveyors, 157
  screw elevators, 157
Minimum hopper opening, 64
Minimum ignition energy, 173
Minimum ignition temperature, 173
Mohr Circle, 15, 36, 37, 55
Moving bed flow, 106

Negative pressure transport, 104
Non-Newtonian slurries, 139
  pressure drop, 142
  rheological models, 139
Numeric-mean diameter, 8

Particle properties, 6
  equivalent diameter, 3
  mean size, 6, 8
  shape, 2, 9
  shape coefficient, 5
  size distribution, 6
  sphericity, 3, 4, 5
  surface area, 2
  surface coefficient, 4
  volume coefficient, 4
Particle-size measurement, 8
Plug flow, 43

Positive pressure pneumatic
  transport, 104, 105
Pressure coefficient, 27, 28
Pressure drops,
  for flow of non-settling slurries, 142
  for flow of settling suspensions, 133
  for pneumatic transport, 113
Pressure transmission in bulk solids, 20
  in bins during discharge, 31
  in bins during filling and storage, 25, 31
  in hoppers during discharge, 35
  in hoppers during filling and storage, 33

Rotary flow conveyors, 166

Saltation, 105
Saltation velocity, 105
Sauter diameter, 8
Screw conveyors, 157
Screw elevators, 157
Silo design,
  funnel flow, 51, 73
  mass flow, 51, 52
Sliding bed flow, 135
Sources of ignition, 171
Soviet CH-302-65, 26
St classification, 174, 176
Standpipes, 93, 97
Star conveyors, 166
Suppressant, 178
Surface-volume diameter, 8

Transition velocity, 135

Unconfined yield strength, 54
Upper explosion limit, 170

van der Waals' forces, 9, 11
Vertical pneumatic transport, 106

Wall yield locus, 70
Weight mean diameter, 8